Ocean Energy: A Formula Handbook

N.B. Singh

DEDICATION

To Nature,

I dedicate this book to you, the source of all life. You are my inspiration, my teacher, and my friend.

Thank you for teaching me about the beauty of the world around me. Thank you for showing me the power of the natural world. Thank you for giving me a sense of peace and tranquillity.

I promise to do my part to protect you and your many wonders. I will teach my children about the importance of conservation and sustainability. I will work to make the world a better place for all living things.

Thank you for everything, Nature.

With love,

N.B Singh

Contents

PREFACE

Welcome to *Ocean Energy: A Formula Handbook*. This handbook is designed as a comprehensive guide for researchers, engineers, and enthusiasts in the field of ocean energy. As the global demand for sustainable energy sources rises, our oceans offer a vast and largely untapped potential.

Scope of the Handbook

This handbook covers a wide array of topics related to ocean energy, ranging from fundamental principles to advanced technologies. It serves as a valuable resource for understanding the theoretical foundations, practical applications, and emerging trends in harnessing energy from the oceans.

Organization of the Handbook

The content is organized into several chapters, each dedicated to a specific aspect of ocean energy. From tidal and wave energy to ocean thermal and seabed resources, each chapter provides insights, formulas, and practical information for readers at various levels of expertise.

Who Can Benefit

Whether you are a student exploring the possibilities of ocean energy, an engineer designing marine energy systems, or a policymaker shaping the future of sustainable energy, this handbook aims to be a one-stop reference guide. The inclusion of mathematical formulas, equations, and real-world examples ensures its applicability across diverse disciplines.

How to Use This Handbook

For optimal use, readers are encouraged to explore chapters based on their specific interests or use the handbook as a sequential guide. The glossary and index provide additional tools for quick reference.

Thank you for embarking on this journey into the world of ocean energy. May this handbook be a valuable companion in your exploration of sustainable and renewable resources beneath the waves.

N.B. Singh

Chapter 1

Introduction

1.1 Background

Oceans, vast and dynamic, conceal a treasure trove of energy. Waves, tides, and temperature gradients possess remarkable potential.

$$\text{Energy} = \text{Mass} \times \text{Gravity} \times \text{Height} \tag{1.1}$$

Waves, with height (H) and wavelength (L), yield energy (E_{wave}) through:

$$E_{\text{wave}} = \frac{1}{2} \times \rho \times g \times H^2 \times L \tag{1.2}$$

Tidal energy, harnessed from amplitude (A) and depth (h), manifests as E_{tide}:

$$E_{\text{tide}} = \frac{1}{2} \times \rho \times g \times A \times h \tag{1.3}$$

Ocean Thermal Energy Conversion (OTEC) exploits temperature differentials (ΔT) with efficiency (η_{OTEC}) given by:

$$\eta_{\text{OTEC}} = \frac{T_c - T_w}{T_c} \times 100\% \tag{1.4}$$

Salinity Gradient Power relies on osmotic pressure (π):

$$\pi = n \times R \times T \times \ln\left(\frac{C_1}{C_2}\right) \tag{1.5}$$

Thermodynamics, encapsulated in $PV = nRT$ and $\Delta G = \Delta H - T\Delta S$, underpin these energy conversions.

$$\Delta G = \Delta H - T\Delta S \tag{1.6}$$

As we embark on unlocking oceanic energy, these formulas propel us toward a sustainable future.

1.2 Scope and Objectives

$$\text{Scope:} \tag{1.7}$$

- Study various forms of ocean energy: tidal (E_{tidal}), wave (E_{wave}), and ocean thermal (E_{OTEC}).

- Investigate methods for harnessing energy from ocean currents ($E_{currents}$), salinity gradients ($E_{salinity}$), and the seabed (E_{seabed}).

- Analyze environmental considerations and impacts: CO_2 emissions, marine life disruption.

- Explore technological innovations: $R_{robotics}$, AI_{ML} applications.

- Assess economic and social impacts: job creation (J_{jobs}), local development (D_{local}).

- Examine case studies: $\Sigma_{studies}$.

$$\text{Objectives:} \tag{1.8}$$

- Develop mathematical models: M_{models} for energy quantification.

- Formulate equations: $E_{efficiency}$ for energy conversion.

- Calculate economic indicators: ROI, C_{kWh}.

- Design experiments: $E_{environment}$ for impact assessment.

- Implement machine learning algorithms: AI_{ML} for optimization.

- Collaborate with interdisciplinary teams: T_{teams} for innovation.

- Engage with policymakers: P_{policy} for governance.

1.3 Significance of Ocean Energy

$$\text{Importance:} \tag{1.9}$$

- Abundant and renewable energy resource: ∞ potential.

- Diverse applications: electricity generation ($E_{electricity}$), desalination ($D_{desalination}$).

- Reduces reliance on fossil fuels: mitigates CO_2 emissions.

- Supports sustainable development: balances $E_{economic}$, $E_{environmental}$ concerns.

$$\text{Challenges:} \tag{1.10}$$

- Technological limitations: efficiency (η) improvements required.

- Environmental impacts: ecosystem disturbance, marine life disruption.

- Economic viability: initial investment ($I_{initial}$), operational costs ($C_{operational}$).

- Policy and regulatory hurdles: permits, licensing.

Chapter 2

Fundamentals of Ocean Energy

2.1   Tidal Energy Basics

Definition: (2.1)

- Tidal energy (E_{tidal}): energy derived from tidal movements.

Key Concepts: (2.2)

- Tidal range (TR): difference in water level between high and low tides.

- Tidal stream velocity (V_{stream}): speed of tidal currents.

- Tidal amplitude (A_{tidal}): maximum height difference between high and low tides.

- Tidal period (T_{tidal}): time between consecutive high (or low) tides.

Equations: (2.3)

- Tidal power (P_{tidal}): $P_{tidal} = \frac{1}{2} \times \rho \times A_{tidal} \times V_{stream}^3 \times$ Area.

- Energy density (ϵ_{tidal}): $\epsilon_{tidal} = \frac{1}{2} \times \rho \times A_{tidal} \times V_{stream}^2$, where ρ is water density.

- Tidal range-to-power ratio ($TRPR$): $TRPR = \frac{TR}{P_{tidal}}$.

2.2   Wave Energy Principles

Waves, nature's rhythmic dance, conceal immense energy potential. Understanding wave energy principles involves embracing both the beauty and the mathematics.

Wave Anatomy:

$$\text{Wave Velocity } (v) = \sqrt{\frac{g}{k}\tanh(kh)}$$

Energy Flux (P):

$$P = \frac{1}{8} \times \rho \times g \times H^2 \times c$$

Wave Power (P_{wave}):

$$P_{\text{wave}} = \frac{1}{16} \times \rho \times g \times H^2 \times c$$

Wave Energy (E_{wave}):

$$E_{\text{wave}} = \frac{1}{64} \times \rho \times g \times H^2 \times c \times T$$

Deep Water Wave Length (L):

$$L = \frac{g \times T^2}{2\pi}$$

Wave Frequency (f):

$$f = \frac{1}{T}$$

Wave Period (T):

$$T = \frac{1}{f}$$

Waves, a culmination of wind energy transfer, harbor potential for harnessing sustainable power. Mathematical insights unveil the harmony between nature's elegance and energy's practicality.

2.3   Ocean Currents

Ocean currents, nature's fluid highways, conceal immense kinetic energy. Navigating the depths requires understanding both the poetry and the mathematics.

Kinetic Energy of Ocean Currents:

$$KE = \frac{1}{2} \times \rho \times A \times v^3$$

Potential Energy from Ocean Currents:

$$PE = \rho \times g \times A \times h$$

Energy Conversion Efficiency (η_{current}):

$$\eta_{\text{current}} = \frac{\text{Extracted Power}}{\text{Available Power}} \times 100\%$$

Underwater Turbine Power (P_{turbine}):

$$P_{\text{turbine}} = \frac{1}{2} \times \rho \times A \times v^3 \times \eta_{\text{current}}$$

Dynamic Positioning Systems (DPS):

$$F_{\text{DPS}} = \rho \times A \times v^2 \times C_d$$

Maintenance Challenges:

$$Cost_{\text{maintenance}} = \frac{\text{Downtime} \times \text{Labor Cost}}{\text{Operating Time}}$$

Integrated Ocean Energy Systems:

$$\text{Synergy Factor} = \frac{\text{Combined Output}}{\text{Sum of Individual Outputs}}$$

Ocean currents, dynamic and powerful, offer a journey into sustainable energy. The fusion of nature's dynamics and mathematical precision propels us toward harnessing this aquatic force.

2.4 Ocean Thermal Energy Conversion (OTEC)

Ocean Thermal Energy Conversion, a marvel of temperature differentials, unlocks a bounty of renewable energy. Let's dive into the heat-powered world of OTEC.

OTEC Process:

$$\Delta T = T_c - T_w$$

Closed-Cycle OTEC Efficiency (η_{closed}):

$$\eta_{\text{closed}} = \frac{\Delta T}{T_c} \times 100\%$$

Open-Cycle OTEC Efficiency (η_{open}):

$$\eta_{\text{open}} = \frac{T_c - T_w}{T_c} \times 100\%$$

Hybrid OTEC Systems:

$$\text{Hybrid Efficiency} = \frac{\text{Closed-Cycle Output} + \text{Open-Cycle Output}}{\text{Total Available Energy}} \times 100\%$$

OTEC Heat Exchanger Area ($A_{\text{heat exchanger}}$):

$$A_{\text{heat exchanger}} = \frac{Q_{\text{heat exchange}}}{U \times \Delta T_{\text{log}}}$$

Materials and Technology Challenges:

$$\text{Material Selection} = \frac{\text{Strength}}{\text{Corrosion Resistance} + \text{Thermal Conductivity}}$$

Economic Feasibility:

$$\text{Levelized Cost of Electricity (LCOE)} = \frac{\text{Total Costs}}{\text{Total Electricity Output}}$$

Global OTEC Projects:

$$\text{Project Scale} = \frac{\text{Installed Capacity}}{\text{Geographical Area}}$$

Ocean Thermal Energy Conversion, a symphony of temperature gradients, beckons us into the realm of sustainable energy exploration. The harmony of thermodynamics and practicality awaits.

2.5   Salinity Gradient Power

Salinity gradient power, derived from the osmotic pressure difference in saltwater, is a promising avenue for sustainable energy. Let's unravel the science and mathematics behind this innovative source.

Osmotic Pressure (π):

$$\pi = n \times R \times T \times \ln\left(\frac{C_1}{C_2}\right)$$

Salinity Gradient Power (P_{salinity}):

$$P_{\text{salinity}} = \pi \times \text{Membrane Surface Area} \times \text{Mass Flow Rate}$$

Energy Efficiency (η_{salinity}):

$$\eta_{\text{salinity}} = \frac{P_{\text{salinity}}}{\text{Available Osmotic Power}} \times 100\%$$

Capacitive Mixing:

$$\text{Energy Density} = \frac{1}{2} \times C \times V^2$$

Concentration Gradient Solar Ponds:

$$\text{Energy Output} = \frac{\text{Heat Extracted}}{\text{Surface Area}}$$

Environmental Impacts:

$$\text{Impact Index} = \frac{\text{Ecological Consequences}}{\text{Energy Production}}$$

Market Trends:

$$\text{Investment Attractiveness} = \text{Return on Investment (ROI)} + \text{Market Growth Rate}$$

Salinity gradient power, a marriage of physics and innovation, holds promise for a sustainable energy future. As we explore its intricacies, the synergy of nature and technology propels us forward.

2.6   Energy from Seabed

Unveiling the hidden energy reservoirs beneath the ocean floor, extracting power from the seabed involves a delicate dance with nature's depths. Let's dive into the practical aspects.

Geothermal Energy:

$$\text{Heat Transfer} = \frac{k \times A \times \Delta T}{d}$$

Methane Hydrates:

$$\text{Methane Release Rate} = \text{Hydrate Decomposition Rate} \times \text{Hydrate Volume}$$

Submarine Volcanic Systems:

$$\text{Heat Output} = \frac{\text{Magma Heat Flux} \times \text{Contact Area}}{\text{Seafloor Area}}$$

Deep-Sea Mining:

$$\text{Resource Extraction Efficiency} = \frac{\text{Valuable Resource}}{\text{Total Resource Deposits}} \times 100\%$$

Submarine Cable Systems:

$$\text{Transmission Efficiency} = \frac{\text{Received Power}}{\text{Transmitted Power}} \times 100\%$$

Ecological Concerns:

$$\text{Biodiversity Impact} = \frac{\text{Species Displacement}}{\text{Total Species in Area}} \times 100\%$$

Technological Advancements:

$$\text{Innovation Index} = \frac{\text{Technological Advancements}}{\text{Implementation Challenges}}$$

Navigating the seabed's energy treasures requires a delicate balance between extraction and conservation. As we plunge into these depths, the synergy of technology and ecology shapes our journey toward sustainable energy solutions.

2.7   Energy from Seabed

Unveiling the hidden energy reservoirs beneath the ocean floor, extracting power from the seabed involves a delicate dance with nature's depths. Let's dive into the practical aspects.

Geothermal Energy:

$$\text{Heat Transfer} = \frac{k \times A \times \Delta T}{d}$$

Methane Hydrates:

$$\text{Methane Release Rate} = \text{Hydrate Decomposition Rate} \times \text{Hydrate Volume}$$

Submarine Volcanic Systems:

$$\text{Heat Output} = \frac{\text{Magma Heat Flux} \times \text{Contact Area}}{\text{Seafloor Area}}$$

Deep-Sea Mining:

$$\text{Resource Extraction Efficiency} = \frac{\text{Valuable Resource}}{\text{Total Resource Deposits}} \times 100\%$$

Submarine Cable Systems:

$$\text{Transmission Efficiency} = \frac{\text{Received Power}}{\text{Transmitted Power}} \times 100\%$$

Ecological Concerns:

$$\text{Biodiversity Impact} = \frac{\text{Species Displacement}}{\text{Total Species in Area}} \times 100\%$$

Technological Advancements:

$$\text{Innovation Index} = \frac{\text{Technological Advancements}}{\text{Implementation Challenges}}$$

Navigating the seabed's energy treasures requires a delicate balance between extraction and conservation. As we plunge into these depths, the synergy of technology and ecology shapes our journey toward sustainable energy solutions.

Chapter 3

Tidal Energy Harvesting

3.1 Tidal Stream Generators

Harnessing the relentless power of tidal streams involves innovative technologies. Let's dive into the dynamics of tidal stream generators with a blend of simplicity and precision.

Tidal Stream Power:

$$P_{\text{tide}} = \frac{1}{2} \times \rho \times A \times v^3$$

Turbine Efficiency (η_{turbine}):

$$\eta_{\text{turbine}} = \frac{P_{\text{turbine}}}{\text{Available Tidal Power}} \times 100\%$$

Blade Tip Speed Ratio (λ):

$$\lambda = \frac{\text{Peripheral Speed}}{\text{Tidal Stream Speed}}$$

Power Coefficient (C_p):

$$C_p = \frac{P_{\text{turbine}}}{\frac{1}{2} \times \rho \times A \times V_{\text{stream}}^3}$$

Tidal Resource Assessment:

$$\text{Tidal Capacity Factor} = \frac{\text{Actual Energy Produced}}{\text{Maximum Potential Energy}} \times 100\%$$

Challenges and Opportunities:

$$\text{Operational Downtime} = \frac{\text{Total Downtime}}{\text{Total Operational Time}} \times 100\%$$

Case Studies:

$$\text{Performance Ratio} = \frac{\text{Actual Output}}{\text{Expected Output}} \times 100\%$$

Tidal stream generators, propellers of the sea, exhibit a dance of efficiency and challenges. As we ride the tides of innovation, these formulas guide our exploration of a sustainable tidal energy future.

3.2   Tidal Range Exploitation

Tapping into the rise and fall of tides, tidal range exploitation presents an array of possibilities. Let's navigate the tidal range dynamics with simplicity and mathematical precision.

Tidal Range Power:

$$P_{\text{tidal}} = \frac{1}{2} \times \rho \times g \times A \times H$$

Tidal Range Efficiency (η_{tidal}):

$$\eta_{\text{tidal}} = \frac{P_{\text{tidal}}}{\text{Available Tidal Power}} \times 100\%$$

Basin Efficiency (η_{basin}):

$$\eta_{\text{basin}} = \frac{\text{Useful Energy Output}}{\text{Tidal Range Potential Energy}} \times 100\%$$

Tidal Lagoons:

$$\text{Energy Storage Capacity} = \text{Reservoir Area} \times \text{Tidal Range} \times \rho \times g$$

Tidal Barrages:

$$\text{Generation Period} = \frac{\text{Time of High Tide - Time of Low Tide}}{2}$$

Tidal Energy Conversion Technologies:

$$\text{Device Efficiency} = \frac{\text{Electrical Power Output}}{\text{Mechanical Power Input}} \times 100\%$$

Challenges in Tidal Range Exploitation:

$$\text{Environmental Impact Index} = \frac{\text{Ecological Consequences}}{\text{Energy Production}} \times 100\%$$

Tidal range exploitation, a rhythmic ballet of efficiency and challenges, unveils the potential within rising and falling tides. As we ride these tidal waves, the mathematical cadence guides us towards a sustainable energy horizon.

3.3 Tidal Lagoons

Tidal lagoons, coastal wonders harnessing tidal rhythms, hold promise for sustainable energy. Let's unravel the dynamics of tidal lagoons with a blend of simplicity and mathematical insight.

Tidal Lagoon Energy Storage Capacity:

$$E_{\text{storage}} = \frac{1}{2} \times \rho \times g \times A \times H^2$$

Tidal Lagoon Energy Output:

$$P_{\text{lagoon}} = \frac{1}{2} \times \rho \times g \times A \times H \times \eta_{\text{lagoon}}$$

Tidal Lagoon Efficiency (η_{lagoon}):

$$\eta_{\text{lagoon}} = \frac{P_{\text{lagoon}}}{\text{Available Tidal Power}} \times 100\%$$

Tidal Lagoon Reservoir Area:

$$A_{\text{reservoir}} = \frac{E_{\text{storage}}}{\rho \times g \times H^2}$$

Tidal Lagoon Generation Period:

$$\text{Generation Period} = \frac{\text{Time of High Tide - Time of Low Tide}}{2}$$

Environmental Impact Index:

$$\text{Impact Index} = \frac{\text{Ecological Consequences}}{\text{Energy Production}} \times 100\%$$

Tidal lagoons, nature's energy reservoirs, beckon us into a dance of efficiency and environmental harmony. As we navigate these coastal marvels, the mathematical rhythms guide us toward unlocking their potential for sustainable energy solutions.

3.4 Tidal Barrages

Tidal barrages, engineering marvels capitalizing on tidal fluctuations, present a dynamic interplay of energy conversion. Let's navigate the principles of tidal barrages with a blend of simplicity and mathematical precision.

Tidal Barrage Energy Output:

$$P_{\text{barrage}} = \frac{1}{2} \times \rho \times g \times A \times H \times \eta_{\text{barrage}}$$

Tidal Barrage Efficiency (η_{barrage}):

$$\eta_{\text{barrage}} = \frac{P_{\text{barrage}}}{\text{Available Tidal Power}} \times 100\%$$

Tidal Barrage Generation Period:

$$\text{Generation Period} = \frac{\text{Time of High Tide - Time of Low Tide}}{2}$$

Tidal Barrage Reservoir Area:

$$A_{\text{reservoir}} = \frac{1}{2} \times \frac{E_{\text{storage}}}{\rho \times g \times H^2}$$

Tidal Barrage Lock Operation:

$$\text{Lock Efficiency} = \frac{\text{Number of Successful Lock Operations}}{\text{Total Number of Attempts}} \times 100\%$$

Sedimentation Management:

$$\text{Dredging Frequency} = \frac{\text{Volume of Sediment Dredged}}{\text{Time Period}}$$

Environmental Impact Index:

$$\text{Impact Index} = \frac{\text{Ecological Consequences}}{\text{Energy Production}} \times 100\%$$

Tidal barrages, orchestrators of tidal energy, invite us to a symphony of efficiency and environmental stewardship. As we navigate the tides with these engineered marvels, the mathematical choreography guides us toward unlocking their potential for sustainable energy solutions.

3.5   Tidal Energy Conversion

Tidal energy conversion, the art of turning tidal forces into electricity, is a symphony of engineering and nature. Let's explore the practical aspects with a rhythmic blend of simplicity and mathematical precision.

Tidal Range Power:

$$P_{\text{tidal}} = \frac{1}{2} \times \rho \times g \times A \times H$$

Tidal Stream Power:

$$P_{\text{stream}} = \frac{1}{2} \times \rho \times A \times v^3$$

Tidal Energy Conversion Efficiency ($\eta_{\text{conversion}}$):

$$\eta_{\text{conversion}} = \frac{\text{Total Electrical Power Output}}{\text{Available Tidal Power}} \times 100\%$$

Combined Tidal Energy:

$$P_{\text{combined}} = P_{\text{tidal}} + P_{\text{stream}}$$

Energy Storage in Batteries:

$$\text{Energy Stored} = \text{Battery Capacity} \times \text{Battery Voltage}$$

Grid Integration:

$$\text{Grid Efficiency} = \frac{\text{Actual Power Delivered to Grid}}{\text{Total Power Produced}} \times 100\%$$

Dynamic Tidal Energy Forecasting:

$$\text{Forecast Accuracy} = \frac{\text{Actual Tidal Energy Output}}{\text{Forecasted Tidal Energy Output}} \times 100\%$$

Tidal energy conversion, a dance of currents and turbines, propels us towards a renewable future. As we navigate these energy waters, the mathematical currents guide us to harness tidal forces with efficiency and precision.

3.6 Challenges in Tidal Energy Extraction

Tidal energy extraction, a promising frontier, is not without its challenges. Let's navigate the hurdles with a pragmatic blend of simplicity and mathematical insight.

Turbine Blade Fatigue:

$$\text{Fatigue Life} = \frac{\text{Number of Stress Cycles endured}}{\text{Stress Cycles per Second}}$$

Tidal Stream Velocity Variability:

$$\text{Coefficient of Variation (CV)} = \frac{\text{Standard Deviation}}{\text{Mean Velocity}} \times 100\%$$

Tidal Range Discrepancies:

$$\text{Range Mismatch} = \frac{\text{Predicted Tidal Range}}{\text{Actual Tidal Range}} \times 100\%$$

Grid Integration Challenges:

$$\text{Power Loss Rate} = \frac{\text{Power Lost during Transmission}}{\text{Total Power Generated}} \times 100\%$$

Environmental Impact Mitigation:

$$\text{Mitigation Index} = \frac{\text{Mitigation Factors}}{\text{Environmental Impact}}$$

Sedimentation and Turbidity Issues:

$$\text{Sedimentation Rate} = \frac{\text{Sediment Deposition}}{\text{Time Period}}$$

Dynamic Tidal Energy Forecasting Accuracy:

$$\text{Forecast Accuracy} = \frac{\text{Actual Tidal Energy Output}}{\text{Forecasted Tidal Energy Output}} \times 100\%$$

Tidal energy extraction, while promising, demands solutions to these challenges. As we confront these hurdles, the mathematical guideposts lead us toward overcoming obstacles and realizing the full potential of tidal energy.

3.7   Case Studies

Let's delve into real-world examples of tidal energy projects, extracting insights with a pragmatic blend of simplicity and mathematical precision.

MeyGen Tidal Project (Scotland):

$$\text{Capacity Factor} = \frac{\text{Actual Energy Produced}}{\text{Maximum Potential Energy}} \times 100\%$$

La Rance Tidal Power Plant (France):

$$\text{Efficiency} = \frac{\text{Actual Power Output}}{\text{Installed Capacity}} \times 100\%$$

Swansea Bay Tidal Lagoon (Wales):

$$\text{Economic Viability} = \text{Net Present Value (NPV)} - \text{Initial Investment}$$

Bay of Fundy Tidal Range (Canada):

$$\text{Power Density} = \frac{\text{Power Output}}{\text{Cross-Sectional Area}}$$

Sihwa Lake Tidal Power Station (South Korea):

$$\text{Annual Energy Production} = \text{Power Output} \times \text{Operating Hours}$$

Pentland Firth and Orkney Waters (UK):

$$\text{Levelized Cost of Energy (LCOE)} = \frac{\text{Total Cost}}{\text{Total Energy Produced}}$$

Tocardo Tidal Power Plant (Netherlands):

$$\text{Performance Ratio} = \frac{\text{Actual Output}}{\text{Expected Output}} \times 100\%$$

Exploring these case studies unveils the intricate interplay of factors influencing the success and challenges of tidal energy projects. As we dissect these real-world examples, the mathematical nuances guide us toward optimizing future endeavors in tidal energy harvesting.

Chapter 4

Wave Energy Conversion

4.1 Wave Energy Converters Overview

Wave energy converters, the architects of harnessing oceanic motion, offer a dynamic interplay of technology and nature. Let's explore the essentials with a rhythmic blend of simplicity and mathematical insight.

Wave Power Extraction:

$$P_{\text{wave}} = \frac{1}{2} \times \rho \times g \times H \times A \times \eta_{\text{wave}}$$

Wave Energy Conversion Efficiency ($\eta_{\textbf{wave}}$):

$$\eta_{\text{wave}} = \frac{P_{\text{converted}}}{P_{\text{wave}}} \times 100\%$$

Oscillating Water Column (OWC):

$$\text{Pressure Head} = \frac{1}{2} \times \rho \times g \times A_{\text{air}} \times H_{\text{wave}}^2$$

Point Absorbers:

$$\text{Buoyant Force} = \rho_{\text{water}} \times g \times V_{\text{displaced}}$$

Attenuation Coefficient (α):

$$\alpha = \frac{\text{Energy Absorbed}}{\text{Wave Intensity}}$$

Linear Generator Power:

$$P_{\text{linear}} = \frac{1}{2} \times \rho \times g \times H_{\text{wave}} \times A_{\text{disp}} \times \omega \times \eta_{\text{generator}}$$

Power Take-Off (PTO) Efficiency ($\eta_{\textbf{PTO}}$):

$$\eta_{\text{PTO}} = \frac{P_{\text{delivered}}}{P_{\text{linear}}} \times 100\%$$

Wave energy converters, orchestrators of the sea's dance, invite us into a realm of efficiency and innovation. As we ride these waves of technology, the mathematical cadence guides us toward unlocking the vast energy potential within oceanic motion.

4.2   Point Absorbers

Point absorbers, resilient energy harvesters riding the waves, epitomize the synergy of technology and oceanic motion. Let's explore their dynamics with a rhythmic blend of simplicity and mathematical precision.

Buoyant Force ($F_{\textbf{buoyant}}$):

$$F_{\text{buoyant}} = \rho_{\text{water}} \times g \times V_{\text{displaced}}$$

Mass-Spring Oscillation:

$$m \cdot a = -k \cdot x + F_{\text{wave}}$$

Power Absorption ($P_{\textbf{absorbed}}$):

$$P_{\text{absorbed}} = \frac{1}{2} \times \rho_{\text{water}} \times g \times A_{\text{disp}} \times H_{\text{wave}} \times \omega \times \eta_{\text{absorber}}$$

Velocity Amplification Factor (VAF):

$$VAF = \frac{\text{Maximum Absorber Velocity}}{\text{Wave Particle Velocity}}$$

Power Conversion Efficiency ($\eta_{\textbf{converter}}$):

$$\eta_{\text{converter}} = \frac{P_{\text{delivered}}}{P_{\text{absorbed}}} \times 100\%$$

Resonance Frequency ($f_{\textbf{resonance}}$):

$$f_{\text{resonance}} = \frac{1}{2\pi} \cdot \sqrt{\frac{k}{m}}$$

Dynamic Response Amplitude (A_{response}):

$$A_{\text{response}} = \frac{F_{\text{wave}}}{m} \cdot \frac{1}{\sqrt{\left(f_{\text{resonance}}^2 - f^2\right)^2 + \left(\frac{f \cdot \gamma}{m}\right)^2}}$$

Point absorbers, surfing the waves of innovation, beckon us into a realm of efficiency and dynamic equilibrium. As we ride these technological waves, the mathematical currents guide us toward optimal energy extraction from the vast expanse of the ocean.

4.3 Attenuators

Attenuators, wave energy dancers on the sea stage, embody the harmony of technology and ocean forces. Let's explore their dynamics with a rhythmic blend of simplicity and mathematical insight.

Wave Power Extraction:

$$P_{\text{wave}} = \frac{1}{2} \times \rho \times g \times H \times A \times \eta_{\text{wave}}$$

Attenuator Power Absorption:

$$P_{\text{absorbed}} = \frac{1}{2} \times \rho \times g \times H_{\text{wave}} \times A_{\text{disp}} \times \omega \times \eta_{\text{attenuator}}$$

Linear Generator Power:

$$P_{\text{linear}} = \frac{1}{2} \times \rho \times g \times H_{\text{wave}} \times A_{\text{disp}} \times \omega \times \eta_{\text{generator}}$$

Wave Attenuation Coefficient (α):

$$\alpha = \frac{\text{Energy Absorbed}}{\text{Wave Intensity}}$$

Conversion Efficiency ($\eta_{\text{converter}}$):

$$\eta_{\text{converter}} = \frac{P_{\text{delivered}}}{P_{\text{absorbed}}} \times 100\%$$

Wave Period (T_{wave}):

$$T_{\text{wave}} = \frac{1}{f_{\text{wave}}}$$

Power Capture Width (PW):

$$PW = \frac{2\pi}{\omega \cdot \alpha}$$

Attenuators, orchestrators of energy ebb and flow, invite us into a realm of efficiency and adaptability. As we navigate these technological waves, the mathematical cadence guides us toward optimal energy extraction from the vast ocean expanse.

4.4   Oscillating Water Columns

Oscillating Water Columns, the rhythmic performers of wave energy, orchestrate a dance with the ocean's breath. Let's dive into their dynamics with a blend of simplicity and mathematical precision.

Pressure Head (P_{head}):

$$P_{head} = \frac{1}{2} \times \rho \times g \times A_{air} \times H_{wave}^2$$

Volume Velocity (Q_{volume}):

$$Q_{volume} = A_{air} \times v_{air}$$

Air Column Resonance Frequency ($f_{resonance}$):

$$f_{resonance} = \frac{c}{4L}$$

Power Conversion Efficiency ($\eta_{converter}$):

$$\eta_{converter} = \frac{P_{delivered}}{P_{head}} \times 100\%$$

Air Column Mass (m_{air}):

$$m_{air} = \frac{\rho_{air} \times A_{air} \times L}{g}$$

Air Turbine Power ($P_{turbine}$):

$$P_{turbine} = \frac{1}{2} \times \rho_{air} \times A_{air} \times v_{air}^3 \times \eta_{turbine}$$

Effective Power Absorption ($P_{effective}$):

$$P_{effective} = \frac{1}{2} \times \rho \times g \times A_{air} \times H_{wave} \times \omega \times \eta_{absorber}$$

Oscillating Water Columns, capturing the ocean's breath, invite us into a world of efficiency and resonance. As we ride these technological waves, the mathematical rhythms guide us toward extracting the vast energy embedded in the ebb and flow of the sea.

4.5   Wave Farms

Wave farms, vast arrays of energy harvesters choreographing the dance of the sea, unfold a symphony of sustainability. Let's delve into their dynamics with a rhythmic blend of simplicity and mathematical precision.

Total Wave Power Absorption:

$$P_{\text{total}} = N \times P_{\text{device}}$$

Wave Farm Efficiency (η_{farm}):

$$\eta_{\text{farm}} = \frac{P_{\text{total}}}{P_{\text{incident}}} \times 100\%$$

Inter-Device Spacing:

$$S_{\text{inter}} = \frac{\sqrt{A_{\text{farm}}}}{N}$$

Array Efficiency Factor (C_{eff}):

$$C_{\text{eff}} = \frac{P_{\text{total}}}{N \times P_{\text{device}}}$$

Power Capture Width Ratio (W_{ratio}):

$$W_{\text{ratio}} = \frac{W_{\text{total}}}{L_{\text{farm}}}$$

Average Power per Unit Length (P_{avg}):

$$P_{\text{avg}} = \frac{P_{\text{total}}}{L_{\text{farm}}}$$

Environmental Impact Index:

$$\text{Impact Index} = \frac{\text{Ecological Consequences}}{\text{Energy Production}} \times 100\%$$

Wave farms, orchestrators of renewable energy, invite us into a realm of efficiency and ecological stewardship. As we navigate these energy seas, the mathematical currents guide us toward optimizing the harmony between technology and the vast oceanic expanse.

4.6 Wave Energy Storage

Wave energy storage, the reservoirs of oceanic power, present a dynamic interplay of capture and release. Let's explore their essence with a rhythmic blend of simplicity and mathematical insight.

Potential Energy Stored:

$$E_{\text{stored}} = \frac{1}{2} \times \rho \times g \times A_{\text{storage}} \times H_{\text{storage}}^2$$

Efficiency of Energy Storage (η_{storage}):

$$\eta_{\text{storage}} = \frac{E_{\text{delivered}}}{E_{\text{stored}}} \times 100\%$$

Wave-to-Wire Efficiency ($\eta_{\text{wave-to-wire}}$):

$$\eta_{\text{wave-to-wire}} = \frac{P_{\text{delivered}}}{P_{\text{incident}}} \times 100\%$$

Storage System Capacity (C_{storage}):

$$C_{\text{storage}} = \frac{E_{\text{stored}}}{V_{\text{storage}}}$$

Wave Energy Release Rate:

$$R_{\text{release}} = \frac{\Delta E_{\text{stored}}}{\Delta t}$$

Charging/Discharging Time ($t_{\text{charge/discharge}}$):

$$t_{\text{charge/discharge}} = \frac{V_{\text{storage}}}{P_{\text{charging/discharging}}}$$

Wave Energy Conversion Efficiency ($\eta_{\text{conversion}}$):

$$\eta_{\text{conversion}} = \frac{P_{\text{delivered}}}{P_{\text{incident}}} \times 100\%$$

Wave energy storage, the reservoirs of power, invite us into a realm of efficient capture and controlled release. As we navigate these energy seas, the mathematical currents guide us toward optimizing the storage and utilization of the vast energy potential within oceanic motion.

4.7   Economic Viability

Economic viability, the financial heartbeat of wave energy ventures, unveils a landscape where sustainability meets fiscal prudence. Let's delve into its dynamics with a rhythmic blend of simplicity and mathematical precision.

Net Present Value (NPV):

$$\text{NPV} = \sum_{t=0}^{T} \frac{R_t}{(1+r)^t} - I_0$$

Internal Rate of Return (IRR):

$$\text{IRR} = \left(\frac{R_0}{I_0}\right)^{1/T} - 1$$

Levelized Cost of Energy (LCOE):

$$\text{LCOE} = \frac{I_0 + \sum_{t=1}^{T} \frac{O \times (1+r)^t}{(1+r)^t}}{\sum_{t=0}^{T} \frac{E \times (1+r)^t}{(1+r)^t}}$$

Payback Period (T_{payback}):

$$T_{\text{payback}} = \frac{I_0}{R_0}$$

Cash Flow at Time t:

$$CF_t = R_t - O_t - D_t$$

Discounted Cash Flow (DCF):

$$\text{DCF} = \sum_{t=0}^{T} \frac{CF_t}{(1+r)^t}$$

Return on Investment (ROI):

$$\text{ROI} = \frac{\text{Total Return}}{\text{Total Investment}} \times 100\%$$

Economic viability, the financial melody of sustainable ventures, invites us to dance between returns and investments. As we navigate these economic currents, the mathematical guideposts lead us toward optimizing the fiscal health of wave energy conversion projects.

Chapter 5

Ocean Currents and Turbines

5.1 Harnessing Ocean Currents

Harnessing ocean currents, the steady rhythm of marine motion, unveils a symphony of energy capture. Let's explore its dynamics with a rhythmic blend of simplicity and mathematical insight.

Kinetic Energy of Ocean Currents:

$$KE_{\text{currents}} = \frac{1}{2} \times \rho \times A_{\text{cross-sectional}} \times v^3$$

Turbine Power Extraction:

$$P_{\text{turbine}} = \frac{1}{2} \times \rho \times A_{\text{turbine}} \times v_{\text{current}}^3 \times \eta_{\text{turbine}}$$

Turbine Efficiency (η_{turbine}):

$$\eta_{\text{turbine}} = \frac{P_{\text{delivered}}}{P_{\text{available}}} \times 100\%$$

Turbine Blade Angle (β):

$$\tan(\beta) = \frac{v_{\text{tangential}}}{v_{\text{current}}}$$

Turbine Rotational Speed (N):

$$N = \frac{v_{\text{tangential}}}{R_{\text{turbine}}}$$

Thrust Force on Turbine Blades:

$$F_{\text{thrust}} = \frac{\dot{m} \cdot (v_{\text{exit}} - v_{\text{current}})}{\Delta t}$$

Turbine Power Coefficient (C_p):

$$C_p = \frac{P_{\text{turbine}}}{\frac{1}{2} \times \rho \times A_{\text{turbine}} \times v_{\text{current}}^3}$$

Harnessing ocean currents, the silent energy dancers of the sea, beckon us into a realm of efficiency and sustainability. As we navigate these currents of opportunity, the mathematical currents guide us toward optimal energy extraction from the vast oceanic motion.

5.2   Underwater Turbines

Underwater turbines, the submerged choreographers of energy extraction, navigate the currents with grace and efficiency. Let's unravel their dynamics with a rhythmic blend of simplicity and mathematical precision.

Turbine Power Extraction:

$$P_{\text{turbine}} = \frac{1}{2} \times \rho \times A_{\text{turbine}} \times v_{\text{current}}^3 \times \eta_{\text{turbine}}$$

Turbine Efficiency (η_{turbine}):

$$\eta_{\text{turbine}} = \frac{P_{\text{delivered}}}{P_{\text{available}}} \times 100\%$$

Turbine Blade Angle (β):

$$\tan(\beta) = \frac{v_{\text{tangential}}}{v_{\text{current}}}$$

Turbine Rotational Speed (N):

$$N = \frac{v_{\text{tangential}}}{R_{\text{turbine}}}$$

Thrust Force on Turbine Blades:

$$F_{\text{thrust}} = \frac{\dot{m} \cdot (v_{\text{exit}} - v_{\text{current}})}{\Delta t}$$

Turbine Power Coefficient (C_p):

$$C_p = \frac{P_{\text{turbine}}}{\frac{1}{2} \times \rho \times A_{\text{turbine}} \times v_{\text{current}}^3}$$

Torque on Turbine Blades:

$$\tau_{\text{turbine}} = \frac{F_{\text{thrust}} \cdot R_{\text{turbine}}}{2}$$

Underwater turbines, silent performers beneath the waves, invite us into a realm of efficiency and sustainable energy. As we dance with these submerged turbines, the mathematical currents guide us toward optimal energy extraction from the ocean's embrace.

5.3 Dynamic Positioning Systems

Dynamic Positioning Systems (DPS), the navigational maestros of ocean current turbines, orchestrate precise alignment for optimal energy extraction. Let's explore their dynamics with a rhythmic blend of simplicity and mathematical insight.

Dynamic Positioning Force:

$$F_{\text{DP}} = m_{\text{turbine}} \times a_{\text{desired}}$$

Thrust Force on Turbine Blades:

$$F_{\text{thrust}} = \frac{\dot{m} \cdot (v_{\text{exit}} - v_{\text{current}})}{\Delta t}$$

Acceleration Required for Dynamic Positioning:

$$a_{\text{desired}} = \frac{v_{\text{exit}} - v_{\text{current}}}{\Delta t}$$

Propulsion Force from Thrust:

$$F_{\text{propulsion}} = \frac{P_{\text{turbine}}}{v_{\text{current}}}$$

Controlled Distance for Dynamic Positioning ($D_{\textbf{control}}$):

$$D_{\text{control}} = \frac{1}{2} \times a_{\text{desired}} \times (\Delta t)^2$$

Dynamic Positioning System Power ($P_{\textbf{DPS}}$):

$$P_{\text{DPS}} = F_{\text{DP}} \times v_{\text{current}}$$

Energy Recovery Efficiency ($\eta_{\textbf{recovery}}$):

$$\eta_{\text{recovery}} = \frac{P_{\text{turbine}}}{P_{\text{DPS}}} \times 100\%$$

Dynamic Positioning Systems, the navigational choreographers beneath the waves, invite us into a realm of precision and energy recovery. As we navigate these technological seas, the mathematical currents guide us toward optimal positioning for harnessing the energy within oceanic motion.

5.4 Maintenance Challenges

Maintenance challenges, the rugged landscapes of turbine longevity, unfold a narrative of resilience and efficiency. Let's navigate through their dynamics with a rhythmic blend of simplicity and mathematical insight.

Turbine Efficiency Over Time ($\eta_{\text{turbine}}(t)$):

$$\eta_{\text{turbine}}(t) = \frac{P_{\text{delivered}}(t)}{P_{\text{available}}(t)} \times 100\%$$

Degradation Rate (γ):

$$\gamma = \frac{\eta_{\text{turbine}}(t_2) - \eta_{\text{turbine}}(t_1)}{t_2 - t_1}$$

Remaining Useful Life (RUL):

$$\text{RUL} = \frac{\eta_{\text{turbine}}(t_{\text{threshold}})}{\gamma}$$

Cost of Maintenance ($C_{\text{maintenance}}$):

$$C_{\text{maintenance}} = \sum_{i=1}^{n} \left(\frac{P_{\text{delivered}}(t_i)}{\eta_{\text{turbine}}(t_i)} \times \text{Cost}_{\text{per kWh}} \right)$$

Mean Time Between Failures (MTBF):

$$\text{MTBF} = \frac{\sum_{i=1}^{n}(t_{\text{operation},i+1} - t_{\text{failure},i})}{n}$$

Availability of Turbine (A_{turbine}):

$$A_{\text{turbine}} = \frac{\text{MTBF}}{\text{MTBF} + \text{Mean Time to Repair (MTTR)}} \times 100\%$$

Maintenance challenges, the rugged trails of turbine existence, invite us into a realm of predictability and cost-effectiveness. As we traverse these challenges, the mathematical currents guide us toward sustaining the longevity and availability of ocean current turbines.

5.5   Integrated Ocean Energy Systems

Integrated Ocean Energy Systems, the harmonious convergence of marine power sources, shape a symphony of sustainable energy. Let's explore their dynamics with a rhythmic blend of simplicity and mathematical insight.

Total Power Output (P_{total}):

$$P_{\text{total}} = P_{\text{turbine}} + P_{\text{wave}} + P_{\text{tidal}} + P_{\text{OTEC}} + P_{\text{salinity}} + P_{\text{seabed}}$$

Energy Conversion Efficiency ($\eta_{\text{conversion}}$):

$$\eta_{\text{conversion}} = \frac{P_{\text{total}}}{P_{\text{incident}}} \times 100\%$$

Synergy Factor ($S_{\textbf{factor}}$):

$$S_{\text{factor}} = \frac{P_{\text{total}}}{\sum_{i=1}^{n} P_{\text{individual, i}}}$$

Power Distribution Ratio ($D_{\textbf{ratio}}$):

$$D_{\text{ratio}} = \frac{P_{\text{individual, i}}}{\sum_{i=1}^{n} P_{\text{individual, i}}}$$

Energy Storage Capacity ($E_{\textbf{storage}}$):

$$E_{\text{storage}} = \frac{1}{2} \times \rho \times g \times A_{\text{storage}} \times H_{\text{storage}}^{2}$$

Optimal Array Configuration:

$$P_{\text{total}} \propto A_{\text{turbine}} + A_{\text{wave}} + A_{\text{tidal}} + A_{\text{OTEC}} + A_{\text{salinity}} + A_{\text{seabed}}$$

Integrated Ocean Energy Systems, the orchestrators of marine energy ballet, invite us into a realm of efficiency and synergy. As we navigate these energy seas, the mathematical currents guide us toward optimal utilization and integration of diverse oceanic power sources.

5.6 Grid Integration

Grid Integration, the seamless merging of marine energy into power networks, orchestrates a dance of reliability and sustainability. Let's navigate through its dynamics with a rhythmic blend of simplicity and mathematical insight.

Power Injection into Grid ($P_{\textbf{grid}}$):

$$P_{\text{grid}} = P_{\text{turbine}} + P_{\text{wave}} + P_{\text{tidal}} + P_{\text{OTEC}} + P_{\text{salinity}} + P_{\text{seabed}}$$

Grid Frequency ($f_{\textbf{grid}}$):

$$f_{\text{grid}} = \frac{1}{T_{\text{grid}}}$$

Synchronization with Grid ($t_{\textbf{synchronize}}$):

$$t_{\text{synchronize}} = \frac{1}{f_{\text{grid}}}$$

Grid Power Factor ($\textbf{PF}_{\textbf{grid}}$):

$$\text{PF}_{\text{grid}} = \cos(\theta)$$

Power Factor Angle (θ):

$$\theta = \arccos(\text{PF}_{\text{grid}})$$

Active Power (P_{active}):

$$P_{\text{active}} = P_{\text{grid}} \times \text{PF}_{\text{grid}}$$

Reactive Power (Q_{grid}):

$$Q_{\text{grid}} = P_{\text{grid}} \times \tan(\theta)$$

Grid Integration, the conductor of marine power within the energy ensemble, invites us into a realm of stability and compatibility. As we synchronize with the grid's heartbeat, the mathematical currents guide us toward harmonizing oceanic energy with the broader power network.

5.7   Regulatory Framework

Regulatory Framework, the guiding compass of marine energy ventures, navigates a course through legal seas. Let's explore its dynamics with a rhythmic blend of simplicity and mathematical insight.

Energy Production License ($L_{\text{production}}$):

$$L_{\text{production}} = \frac{P_{\text{total}}}{P_{\text{capacity}}} \times 100\%$$

Environmental Impact Assessment (EIA):

$$EIA = \frac{\sum_{i=1}^{n} (\text{Impact}_i \times \text{Sensitivity}_i)}{\text{Total Area}}$$

Permit Approval Probability (P_{approval}):

$$P_{\text{approval}} = 1 - \exp\left(-\frac{EIA}{\alpha}\right)$$

Risk Mitigation Factor (α):

$$\alpha = \frac{1}{\text{Risk Tolerance}}$$

Community Acceptance Index (CAI):

$$CAI = \frac{\text{Positive Feedback}}{\text{Negative Feedback} + \text{Neutral Feedback}}$$

Revenue Sharing Mechanism (R_{sharing}):

$$R_{\text{sharing}} = \frac{P_{\text{total}} \times \text{Tax Rate}}{2}$$

Public-Private Partnership Ratio (PPP_{ratio}):

$$PPP_{\text{ratio}} = \frac{\text{Private Investment}}{\text{Public Investment}}$$

Regulatory Framework, the legal navigator of marine energy expeditions, invites us into a realm of compliance and community engagement. As we chart our course through regulatory waters, the mathematical currents guide us toward a sustainable and regulated ocean energy landscape.

Chapter 6

Ocean Thermal Energy Conversion (OTEC)

6.1 OTEC Process

OTEC Process, the heat-driven choreography of oceanic energy, unfolds a dance of sustainability. Let's explore its dynamics with a rhythmic blend of simplicity and mathematical insight.

Temperature Difference (ΔT):

$$\Delta T = T_{\text{surface}} - T_{\text{deep}}$$

Carnot Efficiency (η_{Carnot}):

$$\eta_{\text{Carnot}} = 1 - \frac{T_{\text{deep}}}{T_{\text{surface}}}$$

Power Output (P_{OTEC}):

$$P_{\text{OTEC}} = \eta_{\text{Carnot}} \times Q_{\text{extracted}}$$

Heat Extracted from Ocean ($Q_{\text{extracted}}$):

$$Q_{\text{extracted}} = \rho_{\text{water}} \times C_{\text{water}} \times \dot{m}_{\text{water}} \times \Delta T$$

Working Fluid Selection:

Select working fluid based on temperature range and efficiency.

Cold Water Pipe Design:

$$D_{\text{pipe}} = \sqrt[4]{\frac{4 \times \dot{m}_{\text{water}}}{\pi \times \rho_{\text{water}} \times C_{\text{water}} \times \Delta T}}$$

Environmental Impact (EIA_{OTEC}):

$$EIA_{\text{OTEC}} = \frac{\text{Impact}_{\text{OTEC}} \times \text{Sensitivity}_{\text{OTEC}}}{\text{Total Area}_{\text{OTEC}}}$$

OTEC Process, the ballet of temperature differentials in the oceanic theater, invites us into a realm of efficiency and renewable energy. As we tap into the ocean's thermal embrace, the mathematical currents guide us toward harnessing the vast potential of OTEC.

6.2 Closed-Cycle OTEC

Closed-Cycle OTEC, the thermal tango of sealed systems, orchestrates a symphony of sustainable power. Let's explore its dynamics with a rhythmic blend of simplicity and mathematical insight.

Working Fluid Evaporation ($\dot{m}_{\text{evap}}$):

$$\dot{m}_{\text{evap}} = \frac{Q_{\text{evap}}}{h_{\text{evap}}}$$

Heat Input in Evaporator (Q_{evap}):

$$Q_{\text{evap}} = \rho_{\text{working}} \times \dot{m}_{\text{evap}} \times (h_{\text{evap,out}} - h_{\text{evap,in}})$$

Working Fluid Condensation ($\dot{m}_{\text{cond}}$):

$$\dot{m}_{\text{cond}} = \frac{Q_{\text{cond}}}{h_{\text{cond}}}$$

Heat Rejection in Condenser (Q_{cond}):

$$Q_{\text{cond}} = \rho_{\text{working}} \times \dot{m}_{\text{cond}} \times (h_{\text{cond,in}} - h_{\text{cond,out}})$$

Net Power Output (P_{net}):

$$P_{\text{net}} = \eta_{\text{Carnot}} \times Q_{\text{evap}} - W_{\text{pump}}$$

Pump Work (W_{pump}):

$$W_{\text{pump}} = \rho_{\text{working}} \times \dot{m}_{\text{pump}} \times (h_{\text{pump,out}} - h_{\text{pump,in}})$$

Thermal Efficiency (η_{thermal}):

$$\eta_{\text{thermal}} = \frac{P_{\text{net}}}{Q_{\text{evap}}}$$

Closed-Cycle OTEC, the closed-loop ballet of working fluids, invites us into a realm of efficiency and renewable energy. As we dance with the thermal exchange, the mathematical currents guide us toward harnessing the ocean's thermal richness.

6.3 Open-Cycle OTEC

Open-Cycle OTEC, the dynamic duet of oceanic fluidity, orchestrates a melody of sustainable energy. Let's explore its dynamics with a rhythmic blend of simplicity and mathematical insight.

Warm Seawater Flow Rate ($\dot{m}_{\text{warm}}$):

$$\dot{m}_{\text{warm}} = \frac{Q_{\text{warm}}}{c_{\text{warm}} \times \Delta T_{\text{warm}}}$$

Heat Input in Evaporator (Q_{warm}):

$$Q_{\text{warm}} = \rho_{\text{warm}} \times \dot{m}_{\text{warm}} \times c_{\text{warm}} \times \Delta T_{\text{warm}}$$

Cold Seawater Flow Rate ($\dot{m}_{\text{cold}}$):

$$\dot{m}_{\text{cold}} = \frac{Q_{\text{cold}}}{c_{\text{cold}} \times \Delta T_{\text{cold}}}$$

Heat Rejection in Condenser (Q_{cold}):

$$Q_{\text{cold}} = \rho_{\text{cold}} \times \dot{m}_{\text{cold}} \times c_{\text{cold}} \times \Delta T_{\text{cold}}$$

Net Power Output (P_{net}):

$$P_{\text{net}} = \eta_{\text{Carnot}} \times Q_{\text{warm}} - W_{\text{pump}}$$

Pump Work (W_{pump}):

$$W_{\text{pump}} = \rho_{\text{warm}} \times \dot{m}_{\text{pump}} \times (h_{\text{pump,out}} - h_{\text{pump,in}})$$

Thermal Efficiency (η_{thermal}):

$$\eta_{\text{thermal}} = \frac{P_{\text{net}}}{Q_{\text{warm}}}$$

Open-Cycle OTEC, the oceanic pas de deux of warm and cold seawater, invites us into a realm of efficiency and renewable energy. As we dance with the fluid dynamics, the mathematical currents guide us toward harnessing the ocean's thermal bounty.

6.4   Hybrid OTEC Systems

Hybrid OTEC Systems, the synergy of diverse energy pathways, orchestrates a harmonious composition of sustainable power. Let's explore its dynamics with a rhythmic blend of simplicity and mathematical insight.

Combined OTEC Processes:

Combine Open-Cycle and Closed-Cycle OTEC for improved efficiency.

Heat Input from Warm Seawater (Q_{warm}):

$$Q_{\text{warm}} = \rho_{\text{warm}} \times \dot{m}_{\text{warm}} \times c_{\text{warm}} \times \Delta T_{\text{warm}}$$

Heat Input from Warm Seawater (Q_{evap}):

$$Q_{\text{evap}} = \rho_{\text{evap}} \times \dot{m}_{\text{evap}} \times (h_{\text{evap,out}} - h_{\text{evap,in}})$$

Net Power Output (P_{net}):

$$P_{\text{net}} = \eta_{\text{Carnot}} \times Q_{\text{warm}} + \eta_{\text{closed-cycle}} \times Q_{\text{evap}} - W_{\text{pump}}$$

Pump Work (W_{pump}):

$$W_{\text{pump}} = \rho_{\text{warm}} \times \dot{m}_{\text{pump}} \times (h_{\text{pump,out}} - h_{\text{pump,in}})$$

Thermal Efficiency (η_{thermal}):

$$\eta_{\text{thermal}} = \frac{P_{\text{net}}}{Q_{\text{warm}}}$$

Hybrid OTEC Systems, the orchestrated blend of open and closed cycles, invites us into a realm of efficiency and renewable energy. As we navigate through this energy symphony, the mathematical currents guide us toward optimal utilization of ocean thermal resources.

6.5   Materials and Technology Challenges

Materials and Technology Challenges, the hurdles in the oceanic marathon, unveil the roadmap to sustainable solutions. Let's explore its dynamics with a rhythmic blend of simplicity and mathematical insight.

Material Corrosion Rate (CR):

$$CR = \frac{\text{Corroded Thickness} - \text{Initial Thickness}}{\text{Exposure Time}}$$

Material Selection Criteria:

Select materials with low corrosion rates for longevity.

Thermal Conductivity Enhancement (Δk):

$$\Delta k = k_{\text{enhanced}} - k_{\text{base}}$$

Enhanced Heat Transfer Rate ($\dot{Q}_{\text{enhanced}}$):

$$\dot{Q}_{\text{enhanced}} = \Delta k \times A \times \Delta T$$

Material Strength (σ):

$$\sigma = \frac{\text{Force}}{\text{Cross-sectional Area}}$$

Material Fatigue Life (N_{fatigue}):

$$N_{\text{fatigue}} = \frac{\text{Endurance Limit}}{\text{Applied Stress}}$$

Technology Adaptability Index (TAI):

$$TAI = \frac{\text{Technology Adoption Rate}}{\text{Implementation Cost}}$$

Materials and Technology Challenges, the dynamic duo in the quest for efficiency, invite us into a realm of innovation and sustainable progress. As we confront these challenges, the mathematical currents guide us toward robust materials and cutting-edge technologies for harnessing ocean thermal energy.

6.6 Economic Feasibility

Economic Feasibility, the financial compass in the oceanic journey, navigates us towards sustainable investments. Let's explore its dynamics with a rhythmic blend of simplicity and mathematical insight.

Net Present Value (NPV):

$$NPV = \sum_{t=0}^{T} \frac{\text{Net Cash Flow}_t}{(1+r)^t}$$

Internal Rate of Return (IRR):

$$\text{IRR} = \frac{\text{Cash Inflow}}{\text{Cash Outflow}}$$

Levelized Cost of Electricity (LCOE):

$$\text{LCOE} = \frac{\text{Total Costs}}{\text{Total Electricity Output}}$$

Payback Period (PP):

$$\text{PP} = \frac{\text{Initial Investment}}{\text{Annual Net Cash Flow}}$$

Sensitivity Analysis (ΔNPV):

$$\Delta\text{NPV} = \text{NPV}_{\text{Optimistic}} - \text{NPV}_{\text{Pessimistic}}$$

Risk-Adjusted Return on Investment (ROI):

$$\text{ROI} = \frac{\text{Expected Return}}{\text{Risk-Free Rate}}$$

Economic Feasibility, the captain of financial waters, invites us into a realm of cost-effectiveness and sustainable energy. As we set sail with financial calculations, the mathematical currents guide us toward prudent investment decisions for harnessing ocean thermal resources.

6.7   Global OTEC Projects

Global OTEC Projects, the world's energy choreography, showcases diverse initiatives in harnessing ocean thermal energy. Let's explore its dynamics with a rhythmic blend of simplicity and mathematical insight.

Project Capacity Factor ($\text{CF}_{\text{project}}$):

$$\text{CF}_{\text{project}} = \frac{\text{Actual Output}}{\text{Maximum Potential Output}}$$

Energy Production Rate ($\dot{E}_{\text{project}}$):

$$\dot{E}_{\text{project}} = \text{CF}_{\text{project}} \times \text{Installed Capacity} \times \text{Time}$$

Environmental Impact Index (EII):

$$\text{EII} = \frac{\text{Environmental Impact}}{\text{Energy Produced}}$$

Social Acceptance Index (SAI):

$$SAI = \frac{\text{Public Support}}{\text{Concerns Raised}}$$

Cost-Benefit Ratio (CBR$_{\text{project}}$):

$$CBR_{\text{project}} = \frac{\text{Net Present Value}}{\text{Total Investment}}$$

Global OTEC Integration ($\Sigma \dot{E}_{\text{project}}$):

$$\Sigma \dot{E}_{\text{project}} = \sum_{\text{projects}} \dot{E}_{\text{project}}$$

Global OTEC Projects, the ensemble of sustainable endeavors, invite us into a realm of global energy transformation. As we analyze project metrics, the mathematical currents guide us toward understanding the impact and potential of ocean thermal energy on a global scale.

Chapter 7

Salinity Gradient Power

7.1 Principles of Salinity Gradient Power

Principles of Salinity Gradient Power, the dance of ions in the quest for energy, unveils the secrets of harnessing osmotic pressure. Let's explore its dynamics with a rhythmic blend of simplicity and mathematical insight.

Osmotic Pressure (π):

$$\pi = \frac{n}{V}RT$$

Water Flow ($\dot{V}$):

$$\dot{V} = A \times L \times \frac{\Delta C}{\Delta x}$$

Power Density (P_{density}):

$$P_{\text{density}} = \pi \times \dot{V}$$

Salinity Gradient Power (P_{SGP}):

$$P_{\text{SGP}} = \eta \times P_{\text{density}}$$

Membrane Selectivity (S):

$$S = \frac{\text{Permeability of Water}}{\text{Permeability of Salt}}$$

Efficiency (η):

$$\eta = \frac{P_{\text{SGP}}}{\pi \times \dot{V}}$$

43

Reverse Electrodialysis (RED):

$$2H^+ + 2e^- \longleftrightarrow H_2 \uparrow$$

Principles of Salinity Gradient Power, the ballet of ions, invite us into a realm of renewable energy from the seas. As we decipher the osmotic code, the mathematical currents guide us toward harnessing the power of salinity gradients for sustainable electricity.

7.2   Pressure Retarded Osmosis (PRO)

Pressure Retarded Osmosis (PRO), the engine of osmotic power, orchestrates the controlled release of energy from salinity gradients. Let's explore its dynamics with a rhythmic blend of simplicity and mathematical insight.

Osmotic Pressure Difference ($\Delta\pi$):

$$\Delta\pi = \pi_{\text{high}} - \pi_{\text{low}}$$

Water Flux ($\dot{V}_{\text{water}}$):

$$\dot{V}_{\text{water}} = A_{\text{membrane}} \times L_{\text{membrane}} \times S \times \Delta\pi$$

Power Density (P_{density}):

$$P_{\text{density}} = \dot{V}_{\text{water}} \times (\pi_{\text{high}} + \pi_{\text{low}})$$

PRO Power (P_{PRO}):

$$P_{\text{PRO}} = \eta_{\text{PRO}} \times P_{\text{density}}$$

PRO Efficiency (η_{PRO}):

$$\eta_{\text{PRO}} = \frac{P_{\text{PRO}}}{\dot{V}_{\text{water}} \times (\pi_{\text{high}} + \pi_{\text{low}})}$$

PRO Membrane Selectivity (S):

$$S = \frac{\text{Permeability of Water}}{\text{Permeability of Salt}}$$

Pressure Retarded Osmosis (PRO), the conductor of salinity gradients, invites us into a realm of controlled energy extraction. As we tune into osmotic symphony, the mathematical currents guide us toward efficient utilization of PRO for sustainable power generation.

7.3 Capacitive Mixing

Capacitive Mixing, the synergy of charged particles, unlocks a unique avenue for extracting energy from salinity gradients. Let's explore its dynamics with a rhythmic blend of simplicity and mathematical insight.

Capacitance (C):

$$C = \frac{\varepsilon \varepsilon_0 A}{d}$$

Voltage (V):

$$V = \frac{Q}{C}$$

Energy Density (E_{density}):

$$E_{\text{density}} = \frac{1}{2} C V^2$$

Capacitive Mixing Power (P_{CM}):

$$P_{\text{CM}} = \eta_{\text{CM}} \times E_{\text{density}}$$

CM Efficiency (η_{CM}):

$$\eta_{\text{CM}} = \frac{P_{\text{CM}}}{E_{\text{density}}}$$

Ion Selectivity (S):

$$S = \frac{\text{Ion Mobility}}{\text{Ion Diffusivity}}$$

Capacitive Mixing, the dance of charges, invites us into a realm of efficient energy extraction. As we harness the power of capacitive mixing, the mathematical currents guide us toward sustainable and innovative solutions for salinity gradient power.

7.4 Concentration Gradient Solar Ponds

Concentration Gradient Solar Ponds, the sunlit reservoirs of potential, harness the power of salinity gradients through solar evaporation. Let's explore its dynamics with a rhythmic blend of simplicity and mathematical insight.

Evaporation Rate ($\dot{m}_{\text{evap}}$):

$$\dot{m}_{\text{evap}} = \alpha A_{\text{pond}} \times I_{\text{sun}}$$

Salt Diffusion ($\dot{m}_{\text{diff}}$):

$$\dot{m}_{\text{diff}} = D \times A_{\text{pond}} \times \frac{\Delta C}{\Delta x}$$

Net Salt Extraction Rate ($\dot{m}_{\text{extract}}$):

$$\dot{m}_{\text{extract}} = \dot{m}_{\text{evap}} - \dot{m}_{\text{diff}}$$

Power Density (P_{density}):

$$P_{\text{density}} = \dot{m}_{\text{extract}} \times \Delta H$$

Efficiency (η):

$$\eta = \frac{P_{\text{density}}}{\dot{m}_{\text{evap}} \times \Delta H}$$

Concentration Gradient Solar Ponds, the alchemy of sunlight and salt, invite us into a realm of renewable energy. As we capture the essence of solar evaporation, the mathematical currents guide us toward efficient and sustainable salinity gradient power extraction.

7.5   Environmental Impacts

Environmental Impacts, the eco-balance of salinity gradient power, navigates us through the sustainability challenges. Let's explore its dynamics with a rhythmic blend of simplicity and mathematical insight.

Salinity Discharge ($\dot{S}_{\text{discharge}}$):

$$\dot{S}_{\text{discharge}} = \frac{\dot{m}_{\text{diff}}}{A_{\text{outflow}}}$$

Temperature Change (ΔT):

$$\Delta T = \frac{\dot{Q}_{\text{heat}}}{m_{\text{water}} \times c_{\text{water}}}$$

Chemical Oxygen Demand (COD):

$$\text{COD} = \frac{\text{Mass of Oxygen Consumed}}{\text{Mass of Water Sample}}$$

Biological Oxygen Demand (BOD):

$$\text{BOD} = \frac{\text{Mass of Oxygen Consumed by Microorganisms}}{\text{Mass of Water Sample}}$$

Ecological Footprint (EF):

$$\text{EF} = \frac{\text{Total Impact of Activities}}{\text{Biologically Productive Area}}$$

Eutrophication Potential (EP$_{\text{eutro}}$):

$$EP_{\text{eutro}} = \frac{\text{Mass of Phosphorus Released}}{\text{Mass of Water Sample}}$$

Environmental Impacts, the guardians of nature, invite us into a realm of responsible energy extraction. As we assess the ecological footprint, the mathematical currents guide us toward minimizing environmental consequences for sustainable salinity gradient power.

7.6 Market Trends

Market Trends, the heartbeat of progress, unravels the dynamics of salinity gradient power in the evolving energy landscape. Let's explore its currents with a rhythmic blend of simplicity and mathematical insight.

Global Energy Demand (E_{global}):

$$E_{\text{global}} = \sum E_{\text{source}}$$

Renewable Energy Share (RES):

$$RES = \frac{E_{\text{renewable}}}{E_{\text{global}}}$$

Salinity Gradient Power Contribution (SGP$_{\text{contribution}}$):

$$SGP_{\text{contribution}} = \frac{E_{\text{SGP}}}{E_{\text{renewable}}}$$

Investment Trends (I_{trend}):

$$I_{\text{trend}} = \frac{\Delta I_{\text{SGP}}}{\Delta t}$$

Cost of Energy (COE):

$$COE = \frac{C_{\text{total}}}{E_{\text{produced}}}$$

Return on Investment (ROI):

$$ROI = \frac{\text{Net Profit}}{\text{Investment}}$$

Innovation Index (II):

$$II = \frac{\text{Number of Patents}}{\text{Number of Years}}$$

Market Trends, the compass for energy investors, invite us into a realm of strategic decision-making. As we gauge the contribution of salinity gradient power, the mathematical currents guide us toward harnessing market dynamics for sustainable energy solutions.

7.7   Future Prospects

Future Prospects, the unfolding horizon of possibilities, envisions the trajectory of salinity gradient power in the energy landscape. Let's explore its potential with a rhythmic blend of simplicity and mathematical insight.

Technological Advancements (TA):

$$TA = \frac{\text{Number of Innovations}}{\text{Number of Years}}$$

Efficiency Improvements (EI):

$$EI = \frac{\Delta \eta_{\text{SGP}}}{\Delta t}$$

Cost Reduction Strategies (CRS):

$$CRS = \frac{\Delta C_{\text{total}}}{\Delta t}$$

Market Penetration Rate (MPR):

$$MPR = \frac{\Delta E_{\text{SGP}}}{\Delta E_{\text{renewable}}}$$

Environmental Sustainability Index (ESI):

$$ESI = \frac{\text{Environmental Impact}}{\text{Energy Produced}}$$

Policy Support (PS):

$$PS = \frac{\text{Government Initiatives}}{\text{Regulatory Barriers}}$$

Global Adoption Rate (GAR):

$$GAR = \frac{\text{Number of Deployments}}{\text{Number of Countries}}$$

Future Prospects, the compass for sustainable energy evolution, invite us into a realm of continuous improvement. As we chart the course for salinity gradient power, the mathematical currents guide us toward a future of energy resilience and innovation.

Chapter 8

Energy from Seabed

8.1 Geothermal Energy

Geothermal Energy from the Seabed, the hidden powerhouse beneath the waves, taps into the Earth's thermal reservoirs. Let's dive into its depths with a rhythmic blend of simplicity and mathematical insight.

Heat Transfer Rate ($\dot{Q}$):

$$\dot{Q} = -k \cdot A \cdot \frac{\Delta T}{\Delta x}$$

Geothermal Heat Flux (q):

$$q = \frac{\dot{Q}}{A}$$

Temperature Gradient (∇T):

$$\nabla T = \frac{\Delta T}{\Delta x}$$

Thermal Conductivity (k):

$$k = \frac{\dot{Q} \cdot \Delta x}{A \cdot \Delta T}$$

Seabed Temperature (T_{seabed}):

$$T_{\text{seabed}} = T_{\text{surface}} + \nabla T \cdot H$$

Heat Extraction Rate ($\dot{Q}_{\text{extract}}$):

$$\dot{Q}_{\text{extract}} = q \cdot A_{\text{extract}}$$

Efficiency ($\eta_{\text{geothermal}}$):

$$\eta_{\text{geothermal}} = \frac{\dot{Q}_{\text{extract}}}{\dot{Q}}$$

Geothermal Energy from the Seabed, the thermal symphony beneath the waves, invites us into a realm of sustainable power extraction. As we unravel the mathematical currents, we explore the depths of geothermal potential for a resilient energy future.

8.2   Methane Hydrates

Methane Hydrates from the Seabed, the icy reservoirs of energy waiting to be unlocked, offer a glimpse into the future of sustainable fuel. Let's dive into its depths with a rhythmic blend of simplicity and mathematical insight.

Methane Hydrate Formation ($CH_4 \cdot 6H_2O$):

$$CH_4 + 6H_2O \rightleftharpoons CH_4 \cdot 6H_2O$$

Methane Release Rate ($\dot{m}_{\text{methane}}$):

$$\dot{m}_{\text{methane}} = k \cdot A \cdot \frac{\Delta P}{\Delta t}$$

Pressure Change (ΔP):

$$\Delta P = \rho_{\text{water}} \cdot g \cdot H$$

Gas Hydrate Dissociation Depth ($H_{\text{dissociation}}$):

$$H_{\text{dissociation}} = \frac{\Delta P}{\rho_{\text{hydrate}} \cdot g}$$

Methane Extraction Efficiency ($\eta_{\text{extraction}}$):

$$\eta_{\text{extraction}} = \frac{\dot{m}_{\text{methane}}}{\dot{m}_{\text{total}}}$$

Energy Content of Methane (E_{methane}):

$$E_{\text{methane}} = \dot{m}_{\text{methane}} \cdot \Delta H_{\text{combustion}}$$

Seabed Methane Reserves (R_{methane}):

$$R_{\text{methane}} = A_{\text{hydrate}} \cdot H_{\text{dissociation}}$$

Methane Hydrates from the Seabed, the frozen treasures of energy, invite us into a realm of potential fuel sources. As we unlock the mathematical currents, we explore the depths of methane hydrate extraction for a sustainable energy future.

8.3 Submarine Volcanic Systems

Submarine Volcanic Systems from the Seabed, the fiery furnaces beneath the waves, offer a unique window into harnessing geothermal power. Let's descend into their depths with a rhythmic blend of simplicity and mathematical insight.

Heat Transfer Rate ($\dot{Q}_{\text{volcano}}$):

$$\dot{Q}_{\text{volcano}} = k \cdot A \cdot \frac{\Delta T}{\Delta x}$$

Volcanic Heat Flux (q_{volcano}):

$$q_{\text{volcano}} = \frac{\dot{Q}_{\text{volcano}}}{A}$$

Temperature Gradient ($\nabla T_{\text{volcano}}$):

$$\nabla T_{\text{volcano}} = \frac{\Delta T}{\Delta x}$$

Thermal Conductivity (k_{volcano}):

$$k_{\text{volcano}} = \frac{\dot{Q}_{\text{volcano}} \cdot \Delta x}{A \cdot \Delta T}$$

Seabed Temperature (T_{seabed}):

$$T_{\text{seabed}} = T_{\text{volcano}} + \nabla T_{\text{volcano}} \cdot H_{\text{volcano}}$$

Power Generation ($P_{\text{generation}}$):

$$P_{\text{generation}} = \eta_{\text{conversion}} \cdot \dot{Q}_{\text{volcano}}$$

Energy Potential ($E_{\text{potential}}$):

$$E_{\text{potential}} = P_{\text{generation}} \cdot t_{\text{operation}}$$

Submarine Volcanic Systems from the Seabed, the molten engines of energy, invite us into a realm of continuous power generation. As we navigate the mathematical currents, we explore the depths of geothermal potential for a resilient and sustainable energy future.

8.4   Deep-Sea Mining

Deep-Sea Mining from the Seabed, the subaquatic treasure hunt for valuable resources, unveils a world of opportunities. Let's plunge into the depths with a rhythmic blend of simplicity and mathematical insight.

Resource Extraction Rate ($\dot{m}_{\mathbf{mining}}$):

$$\dot{m}_{\text{mining}} = \rho_{\text{ore}} \cdot A_{\text{mining}} \cdot v_{\text{flow}}$$

Mining Efficiency ($\eta_{\mathbf{mining}}$):

$$\eta_{\text{mining}} = \frac{\dot{m}_{\text{mining}}}{\dot{m}_{\text{total}}}$$

Energy Consumption Rate ($\dot{E}_{\mathbf{consumption}}$):

$$\dot{E}_{\text{consumption}} = P_{\text{consumption}} \cdot t_{\text{operation}}$$

Net Energy Gain ($\Delta E_{\mathbf{net}}$):

$$\Delta E_{\text{net}} = E_{\text{recovered}} - \dot{E}_{\text{consumption}}$$

Ore Value ($V_{\mathbf{ore}}$):

$$V_{\text{ore}} = \text{Market Price} \cdot \dot{m}_{\text{mining}}$$

Economic Viability (EV):

$$\text{EV} = \frac{\Delta E_{\text{net}}}{V_{\text{ore}}}$$

Environmental Impact Index (EII):

$$\text{EII} = \frac{\text{Environmental Impact}}{\dot{m}_{\text{mining}}}$$

Deep-Sea Mining from the Seabed, the aquatic quest for resources, beckons us into a realm of sustainable resource extraction. As we navigate the mathematical currents, we explore the depths of economic viability and environmental considerations for a balanced future.

8.5   Submarine Cable Systems

Submarine Cable Systems from the Seabed, the silent power highways beneath the waves, unravel a network of energy transportation. Let's plunge into the depths with a rhythmic blend of simplicity and mathematical insight.

Power Transmission ($P_{\text{transmission}}$):

$$P_{\text{transmission}} = V_{\text{transmission}} \cdot I_{\text{transmission}}$$

Transmission Efficiency ($\eta_{\text{transmission}}$):

$$\eta_{\text{transmission}} = \frac{P_{\text{received}}}{P_{\text{transmitted}}}$$

Voltage Drop (ΔV_{drop}):

$$\Delta V_{\text{drop}} = I_{\text{transmission}} \cdot R_{\text{cable}}$$

Power Loss (P_{loss}):

$$P_{\text{loss}} = I_{\text{transmission}}^2 \cdot R_{\text{cable}}$$

Cable Heating (ΔT_{cable}):

$$\Delta T_{\text{cable}} = I_{\text{transmission}}^2 \cdot R_{\text{cable}} \cdot t_{\text{operation}}$$

Material Selection (M_{material}):

$$M_{\text{material}} = \frac{\Delta T_{\text{cable}}}{\text{Material Specific Heat}}$$

Environmental Impact (EI_{cable}):

$$\text{EI}_{\text{cable}} = \frac{P_{\text{loss}} \cdot t_{\text{operation}}}{\text{Cable Length}}$$

Submarine Cable Systems from the Seabed, the conduits of energy transfer, invite us into a realm of efficient power transmission. As we navigate the mathematical currents, we explore the depths of transmission efficiency and environmental considerations for a connected and sustainable future.

8.6 Ecological Concerns

Ecological Concerns from Seabed Energy, the guardianship of marine ecosystems, illuminates the delicate balance between progress and preservation. Let's delve into the depths with a rhythmic blend of simplicity and mathematical insight.

Biodiversity Index ($\text{BI}_{\text{seabed}}$):

$$\text{BI}_{\text{seabed}} = \frac{\text{Number of Species}}{\text{Area of Impact}}$$

Ecosystem Resilience (ER_{seabed}):

$$ER_{seabed} = \frac{\text{Recovery Rate}}{\text{Impact Rate}}$$

Environmental Risk (ER_{total}):

$$ER_{total} = BI_{seabed} \cdot ER_{seabed}$$

Habitat Disruption (HD_{seabed}):

$$HD_{seabed} = \frac{\text{Area of Impact}}{\text{Total Habitat Area}}$$

Acoustic Impact (AI_{seabed}):

$$AI_{seabed} = \frac{\text{Sound Pressure Level}}{\text{Reference Sound Level}}$$

Mitigation Strategies (MS_{seabed}):

$$MS_{seabed} = \frac{\text{Effectiveness}}{\text{Cost of Implementation}}$$

Risk Mitigation Index (RMI_{seabed}):

$$RMI_{seabed} = ER_{total} \cdot HD_{seabed} \cdot AI_{seabed} \cdot MS_{seabed}$$

Ecological Concerns from Seabed Energy, the harmony of progress and protection, beckon us to navigate the mathematical currents for a sustainable and balanced future.

8.7   Technological Advancements

Technological Advancements in Seabed Energy, the surging tide of innovation beneath the waves, unveils a realm of possibilities. Let's dive into the depths with a rhythmic blend of simplicity and mathematical insight.

Conversion Efficiency ($\eta_{conversion}$):

$$\eta_{conversion} = \frac{P_{converted}}{P_{harvested}}$$

Data Transmission Rate ($R_{transmission}$):

$$R_{transmission} = \frac{\text{Data Transferred}}{\text{Time}}$$

Operational Availability ($\text{OA}_{\text{seabed}}$):

$$\text{OA}_{\text{seabed}} = \frac{\text{Total Operational Time}}{\text{Total Time}}$$

Reliability Index ($\text{RI}_{\text{seabed}}$):

$$\text{RI}_{\text{seabed}} = \frac{\text{Mean Time Between Failures}}{\text{Mean Time To Repair}}$$

Robotic Maintenance ($\text{RM}_{\text{seabed}}$):

$$\text{RM}_{\text{seabed}} = \frac{\text{Number of Successful Repairs}}{\text{Total Maintenance Attempts}}$$

Cost of Energy ($\text{COE}_{\text{seabed}}$):

$$\text{COE}_{\text{seabed}} = \frac{\text{Total Cost}}{\text{Total Energy Produced}}$$

Innovation Index ($\text{II}_{\text{seabed}}$):

$$\text{II}_{\text{seabed}} = \text{OA}_{\text{seabed}} \cdot \text{RI}_{\text{seabed}} \cdot \text{RM}_{\text{seabed}} \cdot \text{COE}_{\text{seabed}}$$

Technological Advancements in Seabed Energy, the dance of progress and precision, beckon us to navigate the mathematical currents for an innovative and sustainable future.

Chapter 9

Environmental Considerations

9.1 Impact on Marine Ecosystems

Impact on Marine Ecosystems in Environmental Considerations, the delicate dance of progress and preservation beneath the waves, unveils the intricate connection between technology and nature. Let's plunge into the depths with a rhythmic blend of simplicity and mathematical insight.

Ecosystem Health Index ($\text{EHI}_{\text{marine}}$):

$$\text{EHI}_{\text{marine}} = \frac{\text{Biotic Factors}}{\text{Abiotic Factors}}$$

Biodiversity Loss ($\text{BL}_{\text{marine}}$):

$$\text{BL}_{\text{marine}} = \frac{\text{Initial Biodiversity} - \text{Final Biodiversity}}{\text{Initial Biodiversity}}$$

Population Dynamics ($\text{PD}_{\text{marine}}$):

$$\text{PD}_{\text{marine}} = \frac{\text{Birth Rate} - \text{Death Rate}}{\text{Migration Rate}}$$

Carrying Capacity ($\text{CC}_{\text{marine}}$):

$$\text{CC}_{\text{marine}} = \frac{\text{Total Resources}}{\text{Resources Consumed per Individual}}$$

Habitat Restoration ($\text{HR}_{\text{marine}}$):

$$\text{HR}_{\text{marine}} = \frac{\text{Area Restored}}{\text{Total Degraded Area}}$$

Ecological Footprint (EF_{marine}):

$$EF_{marine} = \frac{\text{Total Resource Consumption}}{\text{Biocapacity}}$$

Sustainability Index (SI_{marine}):

$$SI_{marine} = EHI_{marine} \cdot BL_{marine} \cdot PD_{marine} \cdot CC_{marine} \cdot HR_{marine} \cdot EF_{marine}$$

Impact on Marine Ecosystems in Environmental Considerations, the symphony of balance and impact, beckon us to navigate the mathematical currents for a sustainable coexistence.

9.2   Mitigation Strategies

Mitigation Strategies in Environmental Considerations, the compass guiding us towards a sustainable future, unveils practical measures to counter environmental impacts. Let's embark on the journey with a rhythmic blend of simplicity and mathematical insight.

Mitigation Effectiveness ($ME_{strategies}$):

$$ME_{strategies} = \frac{\text{Reduction in Impact}}{\text{Total Impact}}$$

Cost-Benefit Ratio ($CBR_{strategies}$):

$$CBR_{strategies} = \frac{\text{Monetary Benefits}}{\text{Monetary Costs}}$$

Adaptation Capacity ($AC_{strategies}$):

$$AC_{strategies} = \frac{\text{Resilience Enhancement}}{\text{Adaptation Cost}}$$

Technological Innovation ($TI_{strategies}$):

$$TI_{strategies} = \frac{\text{Number of Innovations}}{\text{Implementation Time}}$$

Social Acceptance ($SA_{strategies}$):

$$SA_{strategies} = \frac{\text{Public Approval Rate}}{\text{Implementation Challenges}}$$

Policy Compliance ($PC_{strategies}$):

$$PC_{strategies} = \frac{\text{Compliance Rate}}{\text{Enforcement Effort}}$$

Overall Effectiveness (OE$_{\text{strategies}}$):

$$OE_{\text{strategies}} = ME_{\text{strategies}} \cdot CBR_{\text{strategies}} \cdot AC_{\text{strategies}} \cdot TI_{\text{strategies}} \cdot SA_{\text{strategies}} \cdot PC_{\text{strategies}}$$

Mitigation Strategies in Environmental Considerations, the tapestry of action and impact, beckon us to navigate the mathematical currents for a resilient and harmonious coexistence.

9.3 Biodiversity Preservation

Biodiversity Preservation in Environmental Considerations, the vibrant tapestry of life within the ecosystems, calls for a harmonious coexistence between progress and protection. Let's journey into the depths with a rhythmic blend of simplicity and mathematical insight.

Species Richness (S_{richness}):

$$S_{\text{richness}} = \sum_{i=1}^{n} \frac{1}{\text{Relative Abundance}_i}$$

Shannon Diversity Index (H_{Shannon}):

$$H_{\text{Shannon}} = -\sum_{i=1}^{n} \left(\frac{\text{Relative Abundance}_i}{\sum_{i=1}^{n} \text{Relative Abundance}_i} \cdot \log_2 \frac{\text{Relative Abundance}_i}{\sum_{i=1}^{n} \text{Relative Abundance}_i} \right)$$

Evenness (E_{species}):

$$E_{\text{species}} = \frac{H_{\text{Shannon}}}{\log_2 S_{\text{richness}}}$$

Community Similarity ($CS_{\text{communities}}$):

$$CS_{\text{communities}} = \frac{2 \cdot \text{Number of Shared Species}}{\text{Total Number of Species}_1 + \text{Total Number of Species}_2}$$

Genetic Diversity (D_{genetic}):

$$D_{\text{genetic}} = 1 - \sum_{i=1}^{n} \left(\frac{\text{Frequency of Allele}_i^2}{\sum_{i=1}^{n} (\text{Frequency of Allele}_i)^2} \right)$$

Conservation Priority Index (CPI_{species}):

$$CPI_{\text{species}} = \frac{\text{Endangered Species} \cdot \text{Habitat Suitability}}{\text{Threat Severity} \cdot \text{Conservation Cost}}$$

Biodiversity Preservation in Environmental Considerations, the dance of variety and vitality, beckons us to navigate the mathematical currents for a sustainable and diverse future.

9.4   Climate Change Resilience

Climate Change Resilience in Environmental Considerations, the battle against the rising tides of change, calls for a strategic blend of adaptation and mitigation. Let's delve into the depths with a rhythmic blend of simplicity and mathematical insight.

Greenhouse Gas Equivalents (GHG_{eq}):

$$GHG_{eq} = \sum_{i=1}^{n} (GHG_i \cdot \text{Global Warming Potential}_i)$$

Climate Vulnerability Index ($CVI_{climate}$):

$$CVI_{climate} = \frac{\text{Climate Exposure} + \text{Sensitivity} - \text{Adaptive Capacity}}{3}$$

Sea Level Rise Impact (SLR_{impact}):

$$SLR_{impact} = SLR_{projected} - SLR_{adaptation}$$

Temperature Anomaly ($TA_{anomaly}$):

$$TA_{anomaly} = \text{Observed Temperature} - \text{Baseline Temperature}$$

Emission Reduction Potential ($ERP_{emission}$):

$$ERP_{emission} = \frac{\text{Baseline Emissions} - \text{Reduced Emissions}}{\text{Baseline Emissions}}$$

Resilience Capacity ($RC_{capacity}$):

$$RC_{capacity} = \frac{\text{Adaptive Capacity} + \text{Social Capital} + \text{Economic Flexibility}}{3}$$

Risk of Extinction ($RoE_{species}$):

$$RoE_{species} = \frac{\text{Population Decline Rate}}{\text{Time to Extinction}}$$

Climate Change Resilience in Environmental Considerations, the choreography of adaptation and fortitude, beckons us to navigate the mathematical currents for a sustainable and resilient future.

9.5   Community Engagement

Community Engagement in Environmental Considerations, the heart of sustainable progress, invites us to weave connections between humanity and the environment. Let's embark on a journey of understanding with a rhythmic blend of simplicity and mathematical insight.

Participation Index ($PI_{\text{community}}$):

$$PI_{\text{community}} = \frac{\text{Number of Engaged Individuals}}{\text{Total Community Population}}$$

Social Capital ($SC_{\text{community}}$):

$$SC_{\text{community}} = \frac{\text{Number of Social Connections}}{\text{Total Possible Connections}}$$

Community Empowerment ($CE_{\text{community}}$):

$$CE_{\text{community}} = \frac{\text{Number of Empowered Individuals}}{\text{Total Community Population}}$$

Environmental Awareness ($EA_{\text{community}}$):

$$EA_{\text{community}} = \frac{\text{Number of Environmentally Aware Individuals}}{\text{Total Community Population}}$$

Collaboration Coefficient ($CC_{\text{community}}$):

$$CC_{\text{community}} = \frac{\text{Number of Collaborative Projects}}{\text{Total Potential Projects}}$$

Equity Ratio ($ER_{\text{community}}$):

$$ER_{\text{community}} = \frac{\text{Distribution of Benefits}}{\text{Distribution of Efforts}}$$

Resilience Quotient ($RQ_{\text{community}}$):

$$RQ_{\text{community}} = \frac{\text{Community Resilience}}{\text{Average Individual Resilience}}$$

Community Engagement in Environmental Considerations, the melody of unity and understanding, beckons us to navigate the mathematical currents for a shared and sustainable future.

9.6 Regulatory Compliance

Regulatory Compliance in Environmental Considerations, the compass of responsible action, guides us through the labyrinth of rules and regulations. Let's navigate this landscape with a rhythmic blend of simplicity and mathematical insight.

Compliance Rate ($CR_{\text{organization}}$):

$$CR_{\text{organization}} = \frac{\text{Number of Compliant Actions}}{\text{Total Actions Undertaken}}$$

Environmental Impact Assessment (EIA_{project}):

$$EIA_{\text{project}} = \sum_{i=1}^{n} (\text{Magnitude of Impact}_i \times \text{Probability of Occurrence}_i)$$

Risk Mitigation Index ($RMI_{\text{organization}}$):

$$RMI_{\text{organization}} = \frac{\text{Effectiveness of Mitigation Strategies}}{\text{Level of Identified Risks}}$$

Legal Compliance Score ($LCS_{\text{organization}}$):

$$LCS_{\text{organization}} = \frac{\text{Number of Adhered Regulations}}{\text{Total Applicable Regulations}}$$

Pollution Prevention Index (PPI_{facility}):

$$PPI_{\text{facility}} = \frac{\text{Amount of Pollution Prevented}}{\text{Potential Pollution Generation}}$$

Environmental Performance Index ($EPI_{\text{organization}}$):

$$EPI_{\text{organization}} = \frac{\text{Overall Environmental Performance}}{\text{Benchmark Environmental Performance}}$$

Compliance Cost Ratio ($CCR_{\text{organization}}$):

$$CCR_{\text{organization}} = \frac{\text{Total Compliance Costs}}{\text{Total Operational Costs}}$$

Regulatory Compliance in Environmental Considerations, the harmonious dance with legal norms, beckons us to navigate the mathematical currents for a compliant and sustainable future.

9.7   Monitoring and Assessment

Monitoring and Assessment in Environmental Considerations, the vigilant guardianship of our surroundings, invites us to decipher the language of data for informed decisions. Let's navigate through this realm with a rhythmic blend of simplicity and mathematical insight.

Pollution Index ($PI_{\text{pollution}}$):

$$PI_{\text{pollution}} = \frac{\text{Pollutant Concentration}}{\text{Maximum Allowable Concentration}}$$

Biodiversity Index ($BI_{\text{biodiversity}}$):

$$BI_{\text{biodiversity}} = \frac{\text{Number of Species}}{\text{Maximum Potential Species}}$$

Air Quality Index (AQI_{air}):

$$AQI_{\text{air}} = \frac{\text{Concentration of Air Pollutants}}{\text{Maximum Permissible Limits}}$$

Water Quality Index (WQI_{water}):

$$WQI_{\text{water}} = \frac{\text{Water Quality Parameters}}{\text{Standard Water Quality Criteria}}$$

Ecological Footprint (EF_{ecology}):

$$EF_{\text{ecology}} = \frac{\text{Total Impact}}{\text{Biocapacity}}$$

Environmental Impact Assessment ($EIA_{\text{assessment}}$):

$$EIA_{\text{assessment}} = \frac{\text{Positive Impacts}}{\text{Negative Impacts}}$$

Carbon Footprint (CF_{carbon}):

$$CF_{\text{carbon}} = \frac{\text{Total Emissions}}{\text{Carbon Offset}}$$

Monitoring and Assessment in Environmental Considerations, the symphony of data and discernment, beckons us to navigate the mathematical currents for a balanced and sustainable future.

Chapter 10

Technological Innovations

10.1 Advanced Materials

Advanced Materials in Technological Innovations, the architects of progress, invite us to explore the frontier of possibilities. Let's dive into this realm with a rhythmic blend of simplicity and mathematical insight.

Material Strength ($S_{\mathbf{material}}$):

$$S_{\text{material}} = \frac{\text{Applied Force}}{\text{Cross-Sectional Area}}$$

Thermal Conductivity ($TC_{\mathbf{material}}$):

$$TC_{\text{material}} = \frac{\text{Heat Transfer Rate}}{\text{Temperature Gradient}}$$

Electrical Resistivity ($ER_{\mathbf{material}}$):

$$ER_{\text{material}} = \frac{\text{Electric Field}}{\text{Current Density}}$$

Flexibility Index ($FI_{\mathbf{material}}$):

$$FI_{\text{material}} = \frac{\text{Deformation}}{\text{Original Length}}$$

Magnetic Susceptibility ($MS_{\mathbf{material}}$):

$$MS_{\text{material}} = \frac{\text{Magnetization}}{\text{Applied Magnetic Field}}$$

Chemical Stability (CS_{material}):

$$CS_{\text{material}} = \frac{\text{Chemical Reaction Rate}}{\text{Concentration of Reactants}}$$

Durability Index (DI_{material}):

$$DI_{\text{material}} = \frac{\text{Wear Resistance}}{\text{Time of Use}}$$

Advanced Materials in Technological Innovations, the canvas of the future, beckon us to decipher the mathematical nuances for a resilient and transformative journey.

10.2   Robotics and Automation

Robotics and Automation in Technological Innovations, the choreographers of efficiency, invite us to dance with the machines. Let's embark on this journey with a rhythmic blend of simplicity and mathematical insight.

Robot Efficiency (RE_{robot}):

$$RE_{\text{robot}} = \frac{\text{Useful Work Done}}{\text{Energy Input}}$$

Automation Precision ($AP_{\text{automation}}$):

$$AP_{\text{automation}} = \frac{\text{Accurate Outputs}}{\text{Total Outputs}}$$

Task Completion Time (TCT_{robot}):

$$TCT_{\text{robot}} = \frac{\text{Total Task Time}}{\text{Number of Tasks}}$$

Sensing Capability (SC_{robot}):

$$SC_{\text{robot}} = \frac{\text{Correct Sensing Events}}{\text{Total Sensing Events}}$$

Autonomous Navigation (AN_{robot}):

$$AN_{\text{robot}} = \frac{\text{Successful Navigation}}{\text{Total Navigation Attempts}}$$

Robotic Accuracy (RA_{robot}):

$$RA_{\text{robot}} = \frac{\text{Accurate Movements}}{\text{Total Movements}}$$

Workplace Safety Index ($WSI_{\text{automation}}$):

$$WSI_{\text{automation}} = \frac{\text{Safe Operations}}{\text{Total Operations}}$$

Robotics and Automation in Technological Innovations, the silent conductors of progress, call us to synchronize with the mathematical rhythms for a harmonious future.

10.3 Data Analytics in Ocean Energy

Data Analytics in Ocean Energy, the navigators of insights, invite us to delve into the depths of data-driven solutions. Let's navigate this realm with a rhythmic blend of simplicity and mathematical insight.

Data Quality Index (DQI_{data}):

$$DQI_{\text{data}} = \frac{\text{High-Quality Data Points}}{\text{Total Data Points}}$$

Predictive Accuracy ($PA_{\text{analytics}}$):

$$PA_{\text{analytics}} = \frac{\text{Correct Predictions}}{\text{Total Predictions}}$$

Data Processing Speed ($DPS_{\text{analytics}}$):

$$DPS_{\text{analytics}} = \frac{\text{Processed Data Volume}}{\text{Processing Time}}$$

Decision-making Confidence ($DMC_{\text{analytics}}$):

$$DMC_{\text{analytics}} = \frac{\text{Confident Decisions}}{\text{Total Decisions}}$$

Energy Forecasting (EF_{data}):

$$EF_{\text{data}} = \frac{\text{Forecasted Energy Output}}{\text{Actual Energy Output}}$$

Data Security Resilience ($DSR_{\text{analytics}}$):

$$DSR_{\text{analytics}} = \frac{\text{Secure Data Transactions}}{\text{Total Data Transactions}}$$

Adaptive Learning ($AL_{\text{analytics}}$):

$$AL_{\text{analytics}} = \frac{\text{Learned Patterns}}{\text{Total Patterns Encountered}}$$

Data Analytics in Ocean Energy, the alchemists of information, beckon us to decrypt the mathematical codes for a data-driven odyssey.

10.4   Machine Learning Applications

Machine Learning Applications in Technological Innovations, the architects of intelligent systems, invite us to explore the realms of data-driven decision-making. Let's dive into this dimension with a rhythmic blend of simplicity and mathematical insight.

Prediction Accuracy (PA_{ML}):

$$PA_{\mathrm{ML}} = \frac{\text{Correct Predictions}}{\text{Total Predictions}}$$

Feature Importance (FI_{ML}):

$$FI_{\mathrm{ML}} = \frac{\text{Impact of Feature}}{\text{Total Feature Impacts}}$$

Algorithm Precision (AP_{ML}):

$$AP_{\mathrm{ML}} = \frac{\text{True Positives}}{\text{True Positives} + \text{False Positives}}$$

Model Interpretability (MI_{ML}):

$$MI_{\mathrm{ML}} = \frac{\text{Interpretable Decisions}}{\text{Total Decisions}}$$

Anomaly Detection Sensitivity (ADS_{ML}):

$$ADS_{\mathrm{ML}} = \frac{\text{Detected Anomalies}}{\text{Total Anomalies}}$$

Training Efficiency (TE_{ML}):

$$TE_{\mathrm{ML}} = \frac{\text{Training Time}}{\text{Model Accuracy}}$$

Adaptability Index (AI_{ML}):

$$AI_{\mathrm{ML}} = \frac{\text{Adaptations Successfully Implemented}}{\text{Total Adaptations}}$$

Machine Learning Applications in Technological Innovations, the maestros of intelligent decision-making, call us to decode the mathematical harmonies for a future shaped by data.

10.5   Smart Grid Technologies

Smart Grid Technologies in Technological Innovations, the conductors of energy intelligence, invite us to navigate the dynamic landscape of efficient power management. Let's explore this domain with a rhythmic blend of simplicity and mathematical insight.

Grid Efficiency Index ($GEI_{\text{smart grid}}$):

$$GEI_{\text{smart grid}} = \frac{\text{Useful Energy Delivered}}{\text{Total Energy Input}}$$

Demand Response Ratio ($DRR_{\text{smart grid}}$):

$$DRR_{\text{smart grid}} = \frac{\text{Responsive Demand}}{\text{Total Demand}}$$

Grid Reliability Score ($GRS_{\text{smart grid}}$):

$$GRS_{\text{smart grid}} = \frac{\text{Reliable Operations Time}}{\text{Total Time}}$$

Renewable Integration Capacity ($RIC_{\text{smart grid}}$):

$$RIC_{\text{smart grid}} = \frac{\text{Integrated Renewable Energy}}{\text{Total Energy Input}}$$

Smart Meter Accuracy ($SMA_{\text{smart grid}}$):

$$SMA_{\text{smart grid}} = \frac{\text{Accurate Readings}}{\text{Total Readings}}$$

Grid Resilience Quotient ($GRQ_{\text{smart grid}}$):

$$GRQ_{\text{smart grid}} = \frac{\text{Resilient Events Handled}}{\text{Total Events}}$$

Energy Theft Detection ($ETD_{\text{smart grid}}$):

$$ETD_{\text{smart grid}} = \frac{\text{Detected Theft Incidents}}{\text{Total Incidents}}$$

Smart Grid Technologies in Technological Innovations, the orchestrators of energy efficiency, call us to synchronize with the mathematical harmonies for a sustainable energy future.

10.6 Cybersecurity Challenges

Navigating the digital seas of Technological Innovations, we encounter formidable Cybersecurity Challenges. In this fast-paced exploration, we unveil the mathematical underpinnings of cyber resilience.

Threat Vector (TV_{cyber}):

$$TV_{\text{cyber}} = \frac{\text{Detected Threats}}{\text{Total Threats}}$$

Vulnerability Index (VI_{cyber}):

$$VI_{\text{cyber}} = \frac{\text{Mitigated Vulnerabilities}}{\text{Total Vulnerabilities}}$$

Intrusion Detection Rate (IDR_{cyber}):

$$IDR_{\text{cyber}} = \frac{\text{Detected Intrusions}}{\text{Total Intrusions}}$$

Encryption Strength (ES_{cyber}):

$$ES_{\text{cyber}} = \frac{\text{Secure Communications}}{\text{Total Communications}}$$

Incident Response Time (IRT_{cyber}):

$$IRT_{\text{cyber}} = \frac{\text{Timely Responses}}{\text{Total Incidents}}$$

Authentication Accuracy (AA_{cyber}):

$$AA_{\text{cyber}} = \frac{\text{Accurate Authentications}}{\text{Total Authentication Attempts}}$$

Firewall Effectiveness (FE_{cyber}):

$$FE_{\text{cyber}} = \frac{\text{Blocked Threats}}{\text{Total Threats}}$$

Cybersecurity Challenges in Technological Innovations demand a vigilant eye on these mathematical measures, ensuring a secure voyage through the digital realm.

10.7 Interdisciplinary Research

Interdisciplinary Research in Technological Innovations acts as the bridge, connecting diverse realms for groundbreaking solutions. Let's navigate this fusion with a blend of simplicity and mathematical insight.

Innovation Index ($II_{\text{interdisciplinary}}$):

$$II_{\text{interdisciplinary}} = \frac{\text{Number of Collaborations}}{\text{Total Research Projects}}$$

Impact Factor ($IF_{\text{interdisciplinary}}$):

$$IF_{\text{interdisciplinary}} = \frac{\text{Citations from Different Fields}}{\text{Total Citations}}$$

Research Diversity Ratio ($RDR_{\text{interdisciplinary}}$):

$$RDR_{\text{interdisciplinary}} = \frac{\text{Number of Disciplines Covered}}{\text{Total Disciplines}}$$

Collaboration Efficiency ($CE_{\text{interdisciplinary}}$):

$$CE_{\text{interdisciplinary}} = \frac{\text{Effective Output}}{\text{Total Collaboration Inputs}}$$

Innovation Resilience Quotient ($IRQ_{\text{interdisciplinary}}$):

$$IRQ_{\text{interdisciplinary}} = \frac{\text{Successful Adaptations}}{\text{Total Innovation Challenges}}$$

Cross-Domain Synthesis ($CDS_{\text{interdisciplinary}}$):

$$CDS_{\text{interdisciplinary}} = \frac{\text{Synthesized Concepts}}{\text{Total Concepts Investigated}}$$

Knowledge Transfer Velocity ($KTV_{\text{interdisciplinary}}$):

$$KTV_{\text{interdisciplinary}} = \frac{\text{Rapid Knowledge Dissemination}}{\text{Total Transfer Duration}}$$

Interdisciplinary Research in Technological Innovations, the nexus of diverse wisdom, beckons us to explore mathematical frontiers for a transformative research landscape.

Chapter 11

Economic and Social Impacts

11.1 Job Creation

Job Creation, a cornerstone of economic and social progress, unfolds with tangible impacts. Let's explore this vital dimension through a blend of simplicity and mathematical insights.

Employment Rate (ER):

$$ER = \frac{\text{Number of Employed Individuals}}{\text{Total Working-Age Population}} \times 100\%$$

Job Growth Rate (JGR):

$$JGR = \frac{\text{Change in Employment}}{\text{Initial Employment}} \times 100\%$$

Labor Force Participation (LFP):

$$LFP = \frac{\text{Labor Force}}{\text{Working-Age Population}} \times 100\%$$

Unemployment Rate (UR):

$$UR = \frac{\text{Number of Unemployed Individuals}}{\text{Total Labor Force}} \times 100\%$$

Job Satisfaction Index (JSI):

$$JSI = \frac{\text{Satisfied Workers}}{\text{Total Workers Surveyed}} \times 100\%$$

Economic Output per Employee (EOE):

$$EOE = \frac{\text{Gross Domestic Product (GDP)}}{\text{Total Employment}}$$

Skill Utilization Efficiency (SUE):

$$SUE = \frac{\text{Utilized Skills}}{\text{Total Required Skills}} \times 100\%$$

Job Creation, a dynamic force, intertwines with mathematical metrics, propelling economies and fostering societal well-being.

11.2   Local Economic Development

Local Economic Development (LED) is the heartbeat of community prosperity. Dive into the essence of LED with a blend of simplicity and mathematical insights.

Local Economic Growth (LEG):

$$LEG = \frac{\text{Change in Local GDP}}{\text{Initial Local GDP}} \times 100\%$$

Employment Multiplier Effect (EME):

$$EME = \frac{\text{Total Jobs Created}}{\text{Direct Jobs Created}}$$

Income Distribution Index (IDI):

$$IDI = \frac{\text{Income of Lower Quartile}}{\text{Income of Upper Quartile}}$$

Business Diversity Ratio (BDR):

$$BDR = \frac{\text{Number of Local Business Types}}{\text{Total Local Businesses}} \times 100\%$$

Entrepreneurship Index (EI):

$$EI = \frac{\text{Number of Startups}}{\text{Total Local Enterprises}} \times 100\%$$

Investment Attraction Potential (IAP):

$$IAP = \frac{\text{Local Infrastructure Quality}}{\text{Investment Barriers}}$$

Local Economic Development, an economic symphony, harmonizes with these mathematical metrics, sculpting vibrant and resilient communities.

11.3 Energy Accessibility

Unleash the power of energy accessibility, a catalyst for socio-economic transformation. Let's delve into the core concepts with a mix of simplicity and mathematical insights.

Energy Affordability Index (*EAI*):

$$EAI = \frac{\text{Household Income}}{\text{Energy Expenditure}} \times 100\%$$

Energy Equity Quotient (*EEQ*):

$$EEQ = \frac{\text{Energy Access in Underserved Areas}}{\text{Total Energy Access}} \times 100\%$$

Energy Utilization Efficiency (*EUE*):

$$EUE = \frac{\text{Useful Energy Output}}{\text{Total Energy Input}} \times 100\%$$

Community Energy Resilience (*CER*):

$$CER = \frac{\text{Local Renewable Energy Capacity}}{\text{Total Energy Demand}} \times 100\%$$

Energy Inclusivity Impact (*EII*):

$$EII = \frac{\text{Energy Access for Vulnerable Populations}}{\text{Total Population}} \times 100\%$$

Energy Security Index (*ESI*):

$$ESI = \frac{\text{Stability of Energy Supply}}{\text{Energy Demand Variability}}$$

Energy accessibility, the heartbeat of progress, unfolds its potential through these mathematical metrics, illuminating pathways to a brighter and inclusive future.

11.4 Social Equity

Unravel the dynamics of social equity, a cornerstone for a harmonious society. Let's navigate through key concepts with simplicity and mathematical insights.

Social Equity Index (*SEI*):

$$SEI = \frac{\text{Access to Resources in Underserved Communities}}{\text{Total Resource Availability}} \times 100\%$$

Equality Diversity Inclusion Quotient ($EDIQ$):

$$EDIQ = \frac{\text{Equal Opportunities}}{\text{Diverse Representation}} \times \text{Inclusion Factor}$$

Equitable Access Ratio (EAR):

$$EAR = \frac{\text{Accessible Facilities in Marginalized Areas}}{\text{Total Facilities}} \times 100\%$$

Inclusive Growth Coefficient (IGC):

$$IGC = \frac{\text{Income Growth in Vulnerable Populations}}{\text{Overall Income Growth}} \times 100\%$$

Social Justice Quotient (SJQ):

$$SJQ = \frac{\text{Fair Resource Distribution}}{\text{Total Resource Pool}} \times 100\%$$

Equity Impact Factor (EIF):

$$EIF = \frac{\text{Impact on Disadvantaged Groups}}{\text{Total Project Impact}} \times 100\%$$

Social equity, the bedrock of societal prosperity, unfolds its essence through these mathematical metrics, fostering an environment of fairness and inclusivity.

11.5 Global Energy Markets

Dive into the dynamic realm of global energy markets, where economic forces shape the energy landscape. Let's unravel the intricacies with a blend of simplicity and mathematical insights.

Energy Price Elasticity (EPE):

$$EPE = \frac{\%\ \text{Change in Quantity Demanded}}{\%\ \text{Change in Energy Price}}$$

Market Concentration Index (MCI):

$$MCI = \frac{\text{Market Share of Top Firms}}{\text{Total Number of Firms}} \times 100\%$$

Global Energy Consumption Growth Rate ($GECGR$):

$$GECGR = \frac{\text{Current Energy Consumption - Previous Energy Consumption}}{\text{Previous Energy Consumption}} \times 100\%$$

Energy Trade Balance (ETB):

$$ETB = \text{Energy Exports} - \text{Energy Imports}$$

Renewable Energy Penetration (REP):

$$REP = \frac{\text{Renewable Energy Production}}{\text{Total Energy Production}} \times 100\%$$

Market Integration Factor (MIF):

$$MIF = \frac{\text{Integration of Regional Energy Markets}}{\text{Total Number of Regions}} \times 100\%$$

Navigating the global energy markets becomes more accessible through these mathematical metrics, providing a comprehensive view of the economic and social impacts.

11.6 Policy and Governance

Explore the dynamic intersection of policy and governance in the energy landscape, blending practical insights with mathematical formulations.

Policy Effectiveness Index (PEI):

$$PEI = \frac{\text{Achieved Policy Outcomes}}{\text{Intended Policy Goals}} \times 100\%$$

Governance Resilience (GR):

$$GR = \frac{\text{Stability of Energy Governance Structures}}{\text{Total Period of Governance}}$$

Policy Adoption Rate (PAR):

$$PAR = \frac{\text{Number of Adopted Policies}}{\text{Total Number of Proposed Policies}} \times 100\%$$

Regulatory Compliance Index (RCI):

$$RCI = \frac{\text{Number of Compliant Entities}}{\text{Total Number of Regulated Entities}} \times 100\%$$

Policy Innovation Quotient (PIQ):

$$PIQ = \frac{\text{Number of Innovative Policies}}{\text{Total Policies in Force}} \times 100\%$$

Governance Adaptability (GA):

$$GA = \frac{\text{Adaptation of Governance Structures}}{\text{Changing Energy Landscape}} \times 100\%$$

Navigating the realm of policy and governance is simplified with these mathematical metrics, providing a comprehensive view of their impact on economic and social dimensions.

11.7   Public Perception

Unveil the dynamics of public perception in the energy landscape, blending practical insights with mathematical formulations.

Public Satisfaction Index (PSI):

$$PSI = \frac{\text{Positive Energy Sentiments}}{\text{Total Public Sentiments}} \times 100\%$$

Perceived Economic Impact (PEI):

$$PEI = \frac{\text{Perceived Economic Benefits}}{\text{Total Perceived Impact}} \times 100\%$$

Trust in Energy Policies (TEP):

$$TEP = \frac{\text{Public Trust in Energy Governance}}{\text{Total Trust Factors}} \times 100\%$$

Social Acceptance Rate (SAR):

$$SAR = \frac{\text{Accepted Energy Projects}}{\text{Total Proposed Projects}} \times 100\%$$

Communication Effectiveness (CE):

$$CE = \frac{\text{Effective Energy Communication Instances}}{\text{Total Communication Attempts}} \times 100\%$$

Public Awareness Score (PAS):

$$PAS = \frac{\text{Energy Awareness Impact}}{\text{Total Awareness Initiatives}} \times 100\%$$

Unravel the complexities of public perception with these mathematical metrics, offering a comprehensive understanding of the social fabric surrounding energy initiatives.

Chapter 12

Case Studies

12.1 Tidal Energy Projects

Dive into the dynamic world of tidal energy projects, combining practical insights with mathematical precision.

Project Efficiency (PE):

$$PE = \frac{\text{Actual Energy Output}}{\text{Predicted Energy Output}} \times 100\%$$

Return on Investment (ROI):

$$ROI = \frac{\text{Net Profit}}{\text{Total Investment}} \times 100\%$$

Tidal Turbine Reliability (TR):

$$TR = \frac{\text{Total Operating Hours}}{\text{Total Hours in Deployment}} \times 100\%$$

Environmental Impact Index (EII):

$$EII = \frac{\text{Positive Environmental Impact Factors}}{\text{Total Environmental Impact Factors}} \times 100\%$$

Community Engagement Level (CEL):

$$CEL = \frac{\text{Community Participation Instances}}{\text{Total Engagement Opportunities}} \times 100\%$$

Cost per Kilowatt-Hour ($CpKWH$):

$$CpKWH = \frac{\text{Total Project Cost}}{\text{Total Energy Output}}$$

Explore these key metrics to unravel the success stories and challenges faced by tidal energy projects, providing a comprehensive view of their impact and sustainability.

12.2   Wave Energy Installations

Immerse yourself in the world of wave energy installations, blending real-world insights with mathematical precision.

Energy Capture Efficiency (ECE):

$$ECE = \frac{\text{Actual Energy Captured}}{\text{Theoretical Maximum Energy}} \times 100\%$$

Cost-Benefit Ratio (CBR):

$$CBR = \frac{\text{Net Present Value of Benefits}}{\text{Net Present Value of Costs}}$$

Wave-to-Electricity Conversion (WEC):

$$WEC = \frac{\text{Wave Energy Converted to Electricity}}{\text{Total Wave Energy Harvested}} \times 100\%$$

Annual Energy Production (AEP):

$$AEP = \frac{\text{Total Energy Produced Annually}}{\text{Installed Capacity}}$$

Structural Integrity Factor (SIF):

$$SIF = \frac{\text{Actual Structural Integrity}}{\text{Design Structural Integrity}} \times 100\%$$

Environmental Impact Quotient (EIQ):

$$EIQ = \frac{\text{Positive Environmental Impact Factors}}{\text{Total Environmental Impact Factors}} \times 100\%$$

Delve into these metrics to uncover the performance and challenges faced by wave energy installations, providing a holistic view of their sustainability and impact.

12.3   Successful OTEC Implementations

Dive into the triumphs of Ocean Thermal Energy Conversion (OTEC) through noteworthy implementations, blending practical wisdom with mathematical foundations.

OTEC Efficiency (η_{OTEC}):

$$\eta_{\text{OTEC}} = \frac{\text{Net Power Output}}{\text{Available Ocean Thermal Energy}} \times 100\%$$

Thermal Efficiency (η_{thermal}):

$$\eta_{\text{thermal}} = \frac{\text{Net Power Output}}{\text{Heat Extracted from Ocean}} \times 100\%$$

OTEC Power Density (PD_{OTEC}):

$$PD_{\text{OTEC}} = \frac{\text{Net Power Output}}{\text{OTEC Plant Footprint}}$$

Cold Water Pipe Length (L_{cwp}):

$$L_{\text{cwp}} = \frac{\text{Volume Flow Rate of Cold Seawater}}{\text{Cross-Sectional Area of Pipe}}$$

Economic Viability Index (EVI):

$$EVI = \frac{\text{Net Present Value}}{\text{Total Capital Investment}}$$

Explore these metrics to unravel the success stories of OTEC implementations, providing insights into efficiency, power density, pipe design, and economic viability.

12.4 Salinity Gradient Power Plants

Immerse yourself in the success stories of Salinity Gradient Power Plants, blending practical wisdom with mathematical foundations.

Power Production Rate (PPR):

$$PPR = \frac{\text{Electrical Power Output}}{\text{Membrane Area}}$$

Salt Rejection (SR):

$$SR = \frac{\text{Salt Concentration Difference}}{\text{Inlet Salt Concentration}} \times 100\%$$

Energy Efficiency (EE):

$$EE = \frac{\text{Electrical Power Output}}{\text{Chemical Energy Available}} \times 100\%$$

Membrane Selectivity (MS):

$$MS = \frac{\text{Permeability of Freshwater}}{\text{Permeability of Saltwater}}$$

Freshwater Production Rate (FPR):

$$FPR = \frac{\text{Volume of Freshwater Produced}}{\text{Membrane Area}}$$

Explore these metrics to unveil the achievements of Salinity Gradient Power Plants, providing insights into power production, salt rejection, energy efficiency, membrane selectivity, and freshwater production.

12.5   Deep-Sea Geothermal Ventures

Dive into the world of Deep-Sea Geothermal Ventures, where practical wisdom meets mathematical rigor.

Heat Extraction Rate (HER):

$$HER = \frac{\text{Geothermal Heat Extracted}}{\text{Time}}$$

Reservoir Temperature ($T_{\text{reservoir}}$):

$$T_{\text{reservoir}} = \frac{\text{Heat Extracted}}{\text{Heat Extraction Rate}}$$

Efficiency (E):

$$E = \frac{\text{Net Power Output}}{\text{Geothermal Heat Extracted}} \times 100\%$$

Reinjection Temperature ($T_{\text{reinjection}}$):

$$T_{\text{reinjection}} = T_{\text{reservoir}} - \frac{\text{Temperature Drop}}{2}$$

Brine Discharge Temperature (T_{brine}):

$$T_{\text{brine}} = T_{\text{reinjection}} + \text{Temperature Drop}$$

Uncover the success metrics of Deep-Sea Geothermal Ventures, examining heat extraction rate, reservoir temperature, efficiency, reinjection temperature, and brine discharge temperature.

12.6   Community-Driven Initiatives

Embark on the journey of Community-Driven Initiatives, where practical wisdom converges with mathematical insights.

Community Engagement Index (CEI):

$$CEI = \frac{\text{Number of Engaged Community Members}}{\text{Total Community Population}} \times 100\%$$

Impact Assessment (IA):

$$IA = \frac{\text{Social \& Economic Impact}}{\text{Investment in Community Programs}}$$

Empowerment Quotient (EQ):

$$EQ = \frac{\text{Number of Skill-Building Programs}}{\text{Total Community Initiatives}} \times 100\%$$

Sustainability Rating (SR):

$$SR = \frac{\text{Environmental \& Social Sustainability}}{\text{Community Well-being Index}} \times 100\%$$

Uncover the success metrics of Community-Driven Initiatives, examining community engagement index, impact assessment, empowerment quotient, and sustainability rating.

12.7   Technological Pilots

Dive into the realm of Technological Pilots, where innovation meets real-world application.

Performance Efficiency Index (PEI):

$$PEI = \frac{\text{Actual Technological Output}}{\text{Theoretical Maximum Output}} \times 100\%$$

Reliability Assessment (RA):

$$RA = \frac{\text{Mean Time Between Failures (MTBF)}}{\text{Mean Time To Repair (MTTR)}}$$

Innovation Quotient (IQ):

$$IQ = \frac{\text{Number of Implemented Innovations}}{\text{Total Technological Features}} \times 100\%$$

Adaptability Score (AS):

$$AS = \frac{\text{Technological Flexibility}}{\text{Market Demand Dynamics}} \times 100\%$$

Unveil the insights of Technological Pilots through performance efficiency, reliability assessment, innovation quotient, and adaptability score, creating a synergy of innovation and reliability.

Chapter 13

Future Trends and Challenges

13.1 Emerging Technologies

Explore the forefront of innovation with Emerging Technologies, shaping the future of our endeavors.

Quantum Computing Potential:

$$Q_{\text{speedup}} = \frac{\text{Time Complexity of Best-known Classical Algorithm}}{\text{Time Complexity of Quantum Algorithm}}$$

Nanotechnology Advancements:

$$S_{\text{surface area}} = \frac{6}{r_{\text{particle}}}$$

Biotechnological Breakthroughs:

$$\text{Genetic Modification Efficiency} = \frac{\text{Desired Trait Expressions}}{\text{Total Modified Organisms}} \times 100\%$$

Neural Network Performance:

$$\text{NN}_{\text{accuracy}} = \frac{\text{Correct Predictions}}{\text{Total Predictions}} \times 100\%$$

Embrace the potential of Quantum Computing, harness the power of Nanotechnology, unlock Biotechnological frontiers, and enhance Neural Network performances in this era of Emerging Technologies.

13.2 International Collaboration

Forge global partnerships for a sustainable future through effective international collaboration.

Collaboration Index:

$$CI = \frac{\text{Number of Collaborative Projects}}{\text{Total Projects Worldwide}} \times 100\%$$

Knowledge Transfer Impact:

$$KTI = \frac{\text{International Research Publications}}{\text{Total Research Publications}} \times 100\%$$

Economic Alliance Strength:

$$EAS = \frac{\text{Total Value of International Energy Agreements}}{\text{Global Economic Output}} \times 100\%$$

Embrace the Collaboration Index, leverage Knowledge Transfer Impact, and strengthen Economic Alliances for a united front in addressing global challenges.

13.3   Policy Evolution

Navigate the future by adapting to dynamic policy landscapes, crucial for sustainable energy development.

Policy Flexibility Index:

$$PFI = \frac{\text{Number of Policy Revisions}}{\text{Total Policy Duration}} \times 100\%$$

Global Policy Influence:

$$GPI = \frac{\text{International Adoption of Policies}}{\text{Total Nations}} \times 100\%$$

Public Perception Alignment:

$$PPA = \frac{\text{Public Support for Policies}}{\text{Total Public Awareness}} \times 100\%$$

Embrace a high Policy Flexibility Index, expand Global Policy Influence, and ensure Public Perception Alignment for effective policy evolution.

13.4   Market Dynamics

Navigating the evolving energy market is crucial for sustained growth and competitiveness.

Market Share Evolution:

$$MSE = \frac{\text{Change in Company Market Share}}{\text{Total Market Share Change}} \times 100\%$$

Adoption Rate Acceleration:

$$ARA = \frac{\text{Increase in Technology Adoption Rate}}{\text{Total Time Period}} \times 100\%$$

Economic Viability Index:

$$EVI = \frac{\text{Profitability Factor}}{\text{Risk Factor}}$$

Stay ahead by monitoring Market Share Evolution, driving Adoption Rate Acceleration, and ensuring a positive Economic Viability Index for sustainable growth.

13.5   Research and Development

Staying at the forefront of innovation through strategic research and development is paramount.

Innovation Index:

$$II = \frac{\text{Number of Patents Filed}}{\text{Research Expenditure}}$$

Collaboration Coefficient:

$$CC = \frac{\text{Number of Collaborative Projects}}{\text{Total Research Initiatives}} \times 100\%$$

Technology Readiness Level (TRL):

$$TRL = \frac{\text{Readiness Level of Technology}}{9} \times 100\%$$

Enhance the Innovation Index, boost the Collaboration Coefficient, and elevate the Technology Readiness Level to drive impactful research and development.

13.6   Adaptation to Climate Change

Adapting to climate change is imperative for sustainability and resilience.

Climate Vulnerability Index (CVI):

$$CVI = \frac{\text{Climate Risk}}{\text{Adaptation Capacity}}$$

Resilience Quotient (RQ):

$$RQ = \frac{\text{Resilience Measures Implemented}}{\text{Climate-Induced Disruptions}} \times 100\%$$

Adaptation Strategies:

- Implement green infrastructure.

- Enhance coastal protection measures.

- Foster community-based adaptation.

Boost the Climate Vulnerability Index, elevate the Resilience Quotient, and implement robust adaptation strategies for a climate-resilient future.

13.7 Resilience and Sustainability

Achieving resilience and sustainability is pivotal for enduring energy solutions.

Energy Resilience:

$$ER = \frac{\text{Energy Output during Disruptions}}{\text{Total Energy Output}} \times 100\%$$

Sustainability Index (SI):

$$SI = \frac{\text{Renewable Energy Output}}{\text{Total Energy Output}} \times 100\%$$

Strategies for Enhancing Resilience and Sustainability:

- Implement smart grid technologies.

- Integrate energy storage solutions.

- Emphasize circular economy principles.

Elevate Energy Resilience (ER) and Sustainability Index (SI) through innovative technologies and eco-friendly practices for a robust and sustainable energy future.

Chapter 14

Conclusion

14.1 Summary of Key Findings

In conclusion, our exploration into ocean and seabed energy has unearthed crucial insights:

Tidal Energy:

$$P = \frac{1}{2}\rho g H^2 T$$

Tidal energy exhibits high predictability, making it a stable source for power generation.

Wave Energy:

$$P_w = \frac{1}{16}\rho g H^2 c$$

Wave energy converters, including point absorbers and attenuators, harness the ocean's dynamic forces efficiently.

Ocean Thermal Energy Conversion (OTEC):

$$COP = \frac{T_h}{T_h - T_c}$$

OTEC systems, whether closed-cycle or open-cycle, offer sustainable energy by leveraging temperature differences.

Salinity Gradient Power:

$$P_{\text{max}} = \frac{1}{2}\rho g H$$

PRO and capacitive mixing present viable methods for extracting energy from salinity gradients.

Seabed Energy:

$$Q = kA\left(\frac{\Delta T}{d}\right)$$

Methane hydrates and geothermal systems demonstrate potential for tapping into the rich energy reservoirs beneath the seabed.

The future of ocean energy lies in embracing diverse technologies and sustainable practices to meet global energy demands resiliently and responsibly.

14.2   Implications for the Future

As we gaze into the future of ocean and seabed energy, several key implications come to light:

Renewable Integration: The seamless integration of ocean energy into existing power grids demands robust technologies and smart grid solutions. Balancing intermittent energy sources requires advanced control systems.

Environmental Stewardship: Mitigating ecological impacts necessitates continual research on eco-friendly materials and innovative technologies. Striking a balance between energy production and marine biodiversity is crucial.

Technological Advancements: Embracing cutting-edge technologies, such as robotics and data analytics, will enhance the efficiency of energy harvesting and reduce maintenance costs.

International Collaboration: Collaborative efforts on a global scale are essential for addressing climate change and ensuring sustainable development. Shared research initiatives and standardized regulations can propel the industry forward.

Adaptation to Climate Change: Designing resilient energy systems capable of withstanding the effects of climate change is imperative. Incorporating climate resilience into infrastructure planning ensures long-term viability.

Market Dynamics: Dynamic market forces require adaptive policies and governance. Understanding global energy market trends enables countries to position themselves strategically in the evolving energy landscape.

Social and Economic Impacts: Fostering job creation, ensuring local economic development, and promoting social equity are vital for the widespread acceptance and success of ocean energy initiatives.

Research and Development: Continuous investment in R&D will unlock new frontiers, addressing challenges and pushing the boundaries of what is possible in ocean and seabed energy exploration.

In essence, navigating these implications will shape a future where ocean energy contributes significantly to a sustainable, resilient, and equitable global energy landscape.

14.3 Call to Action

As we conclude our exploration of ocean and seabed energy, a resounding call to action emerges. The future of sustainable energy relies on collective efforts and immediate initiatives:

Policy Advocacy: Advocate for policies that incentivize the development and implementation of ocean energy projects. Governments and regulatory bodies play a pivotal role in shaping the energy landscape.

Investment and Funding: Encourage private and public investment in ocean energy research and infrastructure. Financial support is crucial for scaling up technologies and realizing the full potential of marine resources.

Public Awareness: Raise awareness about the benefits of ocean and seabed energy. Informed communities can drive demand, influence policies, and contribute to the success of sustainable energy projects.

Education and Research: Invest in education and research programs focused on ocean energy technologies. Training the next generation of scientists and engineers ensures a steady influx of talent into the field.

Global Collaboration: Foster international collaboration to share knowledge, best practices, and resources. Working together on a global scale accelerates advancements and addresses shared challenges.

Environmental Responsibility: Prioritize environmental responsibility in every phase of ocean energy projects. Sustainable practices and technologies are vital for minimizing ecological impacts.

Innovation and Adaptability: Embrace innovation and adaptability in the face of challenges. A culture of continuous improvement and learning will drive the evolution of ocean energy technologies.

This call to action invites stakeholders from all sectors—governments, industries, academia, and communities—to actively participate in building a sustainable and resilient energy future, where the vast potential of our oceans is harnessed for the benefit of generations to come.

www.ingramcontent.com/pod-product-compliance
Lightning Source LLC
Chambersburg PA
CBHW080958290526
45795CB00009B/2992